IntechOpen

Agronomy - Climate Change & Food Security

Edited by Dr. Amanullah

Published in London, United Kingdom

Supporting open minds since 2005

Agronomy - Climate Change & Food Security
http://dx.doi.org/10.5772/intechopen.78102
Edited by Dr. Amanullah

Assistant to the Editor(s): Shaha Khalid

Contributors
Dr. Amanullah, Shaha Khalid, Prashant Kaushik, Navjot Singh Brar, Dinesh Saini, Pablo Gonzalez-De-Santos, Roemi Fernandez, Delia Sepulveda, Eduardo Navas, Manuel Armada, Liliane Tandzi, Charles Shelton Mutengwa, Mustafa Yildiz, Hudaverdi Gurkan, Yasin Ozgen, Nilgun Bayraktar, Huseyin Bulut, Vinay Sharma, Dinesh Kumar Saini, Ashish Kumar, Hari Kesh, Jyoti Chauhan, Navish Kumar Kamboj

First published in London, United Kingdom, 2020 by IntechOpen
IntechOpen is the global imprint of INTECHOPEN LIMITED, registered in England and Wales, registration number: 11086078, 7th floor, 10 Lower Thames Street, London, EC3R 6AF, United Kingdom
Printed in Croatia

British Library Cataloguing-in-Publication Data
A catalogue record for this book is available from the British Library

Additional hard and PDF copies can be obtained from orders@intechopen.com

Agronomy - Climate Change & Food Security
Edited by Dr. Amanullah
p. cm.
Print ISBN 978-1-83881-222-5
Online ISBN 978-1-83881-223-2
eBook (PDF) ISBN 978-1-83881-224-9

For EU product safety concerns:
IN TECH d.o.o., Prolaz Marije Krucifikse Kozulić 3, 51000 Rijeka, Croatia, info@intechopen.com or visit our website at intechopen.com.

Meet the editor

Dr. Amanullah is currently Associate Professor in the Department of Agronomy, Faculty of Crop Production Sciences, University of Agriculture Peshawar, Pakistan. Dr. Amanullah obtained a PhD in Agronomy from the University of Agriculture Peshawar in 2004 and a post doctorate from Dryland Agriculture Institute, WTAMU, Canyon Texas, USA, in 2010. He has published more than twenty books and more than 200 research papers in peer-reviewed journals, including 100 papers in impact factor journals. He has edited three books: *Rice–Technology and Production* (2017), *Nitrogen in Agriculture-Updates* (2018), and *Corn: Production and Human Health in Changing Climate* (2018). Dr. Amanullah has been awarded three Research Productivity Awards by the Pakistan Council for Science and Technology (PCST), Islamabad. He represented Pakistan in the FAO Intergovernmental technical panel on soil of the Global Soil Partnership (2015–2018). Dr. Amanullah also won first prize in the innovative research proposal competition arranged by DICE at the University of Gujarat in 2013–2014.

Contents

Preface

Agronomy is a branch of agricultural science that deals with production of crops and soil management. The production of crops will need to increase by 50 percent over the next few decades to fulfil expected demand as the world population is estimated to increase to 9 billion by 2050. The rapidly growing population and the increase in demand for food, feed and fuel will require sustainable agronomic practices to increase crop productivity. Sustainable agronomic practices are important to improve food security in changing climates. Agronomy is closely related to two of the United Nations' Sustainable Development Goals (SDGs). SDG-2 (Zero Hunger) focuses explicitly on food by seeking to end hunger, achieve food security and improved nutrition, and promote sustainable agriculture. SDG-13 (Climate Change) calls for urgent actions to combat the negative effects of climate change. The main job of agronomists is to introduce a crop production system through which it makes possible the best use of light, heat, water and soil for sustainable and greater production of crops that are more suitable and resilient to the socioeconomic needs of farming communities, like poverty and hunger alleviation, food security, climate change adaptation and environmental protection.

Agronomy: Food Security and Climate Change presents, over six chapters, a comprehensive picture of the importance of agronomy in SDGs. It is designed for students, researchers, technologists and policy makers.

We are thankful to all authors who contributed their valuable chapters to this book. We are also extremely grateful to IntechOpen's Author Service Manager Ms. Romina Rovan for helping us to publish this book in excellent form in the shortest possible time. We owe our sincere thanks and gratitude to our families, whose consistent encouragement and love have been a tremendous impetus for the completion of this book.

Dr. Amanullah
Department of Agronomy,
Faculty of Crop Production Sciences,
University of Agriculture Peshawar,
Pakistan

Chapter 1

Agronomy-Food Security-Climate Change and the Sustainable Development Goals

Amanullah and Shah Khalid

Abstract

Climate change has negative effects on food security, water security, and energy security due to change in extreme events such as floods, droughts, and heat waves, and reduces agricultural productivity. Global demand for food is projected to double by 2050. The rapidly growing population and the increase in demand for food, feed, and fuel will require sustainable agronomic practices to increase crop productivity. To meet the challenge, many advanced agronomic practices have been developed. For example: (1) selection of suitable crops and their varieties that are resistant to biotic stresses, (2) selection of suitable crops and their varieties that are resistant to abiotic stresses (3) selection of suitable cropping system, sustainable intensification. Sustainable agronomic practices are important to improve food security in changing climates. SDG-2 focuses explicitly on food by seeking to "end hunger, achieve food security and promote sustainable agriculture". SDG-1 focuses on poverty reduction, where agriculture have a key role to play. SDG-13 specifically calls for "urgent actions to combat climate change and its impacts." About 45 of the 169 targets are related to SDG-13, which highlights the need to tackle climate change and avert its impacts, particularly on food security, water, energy, and economic development.

Keywords: agronomy, agriculture, sustainable development goals, SDGs, climate change, GHG gases, food security

1. Introduction

The main job of agronomists is introducing such sophisticated production system through which is made possible the best use of light, heat, water, and soil for the crop production. Haman intervention in production and consumption of food and feed for human and their animals is accountable in climate change, which gives rise to other environmental changes like change in biodiversity, carbon and nitrogen cycling, and fresh water supplies [1, 2]. Change in climate may bring positive effects in some part of the world, particularly above about 55° northern altitudes. Change in climate, will further complex to attain food security in developing countries because of the negative impacts of climate change on crop production especially in subtropical and tropical areas [3–7]. There are three major factors responsible for the climate change and their negative impacts. First, several developing countries are exposed to considerable change in rainfall and temperature; according to the IPCC [8] prediction, the tropical and subtropical areas can experience an increase in temperature of 2–5°C. An intensification in extreme events (droughts and floods) in terms of frequency and intensity is also expected [8, 9]. Second, most of the developing countries' economies are sensitive to the direct deleterious effect of climate change because of the major dependence of developing countries' economies on agriculture and due to the higher poverty level [6]. Third, in developing countries most of the people depend on agriculture directly for their food and livelihood and change in climate will have negative impacts on production of crop and food supply. It is clear that increase in production of crops will need to increase by 50% over next few decades to fulfill the expected demand [10] as the world population is expected to increase from 6 billion to 9 billion by 2050 [11]. The current demands of the people from the existing current production technology and cropping system may further raise environmental complexities [12, 13]; for example, increase in chemical fertilizers can increase GHG emission, which in turn can cause climate which is sometimes also called climate forcing and such changes can cause further decline in crop production. Agronomist have two major challenges. First is the production of food on sustainable basis with the changing climate along with reduction in climate forcing factors and secondly, efforts to more efficiently collaborate with other disciplines to enhance the supply of agronomic products both better integrated within the overall context of food security and better tuned to the needs of food security policy formulation.

2. Agronomy and its scope

Agronomy is a wide and dynamic discipline. Agronomy science becomes imperative in Agriculture in the following areas. Identification of proper season for cultivation of wide range of crops, proper sowing methods, weeds control through different techniques, use of chemical fertilizers and organic manures like poultry manure, farmyard manure, green manuring, brown manuring, compost formation, use of bioherbicides, different cropping techniques like intercropping, monocropping, extensive cropping, intensive cropping, storage techniques for different agricultural produces, water management, management of crop under changing climate and other farm management services broaden the scope of agronomy etc. Agronomy also has a strong relation with other agricultural sciences like agriculture chemistry, plant breading and genetics, plant ecology, crop physiology, economics, and biochemistry (**Figures 1** and **2**).

3. Agronomists and their role in agriculture

Like agronomy, an agronomist also has a vast responsibility. These scientists study various crop production problems and work for better soil and crop management to get higher yield; in a broader sense agronomists deal with production of food, feed, fiber for fulfillment of the needs of the growing population by recommending best crop variety, proper sowing time, sowing method, irrigation time and amount, weed control methods, and proper cropping techniques.

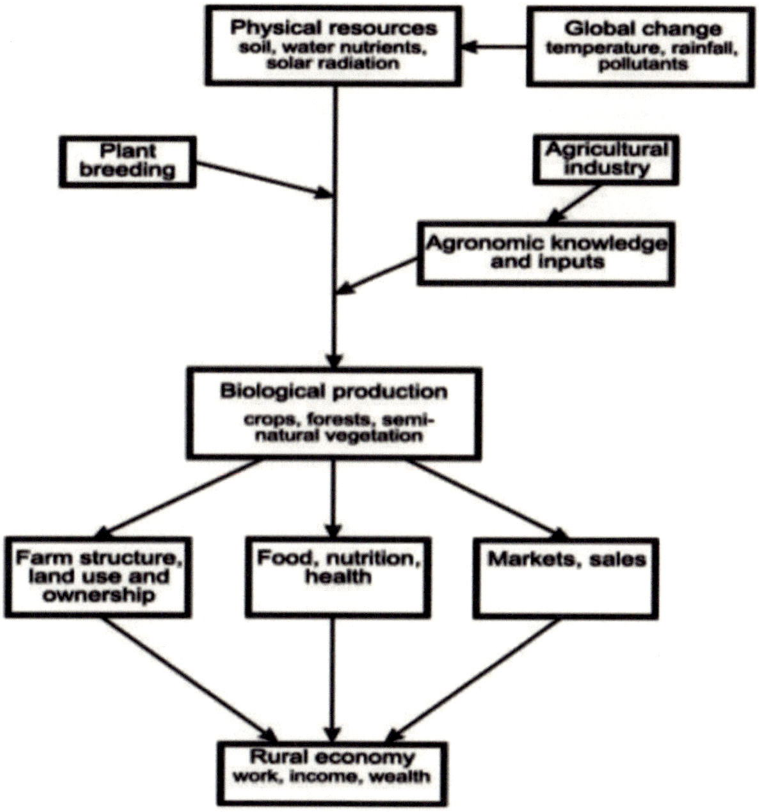

Figure 1.
Flow diagram of physical, biological, economic, and social dimensions of agronomy.

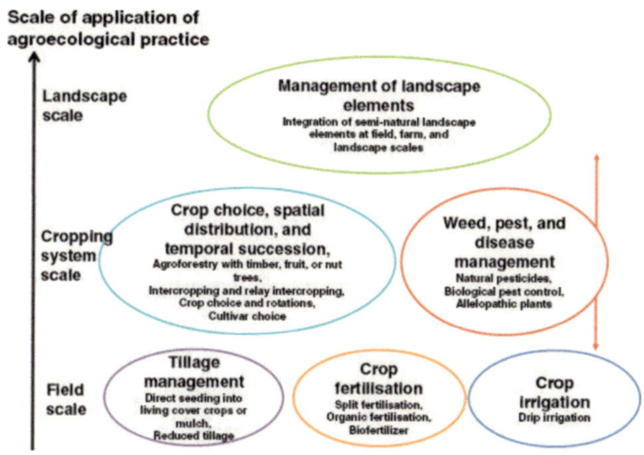

Figure 2.
Various classes of agroecological practices for increasing crop productivity at ranges from field scale to landscape scale.

4. Agronomy and climate change

Enhancement in agronomy and plant breeding has enabled increase in yield of crops over last four decades or more. Yield of many crops, particularly cereal crops like wheat, rice, and maize, has steadily increased in last few decades; this increase is due to improvement in irrigation, fertilizer use, chemical herbicides for weed control and pesticides for pest and disease control, adaptation of new production technology, high-yield varieties, and improvements in crop phenotype from breeding, especially the widespread adoptions of semi-dwarfing genes in cereals [14] has resulted in yield per unit area increase. However, this increase was not similar throughout the world. Yield increase was observed in Europe, America, and Asia, but there was decline in African countries in crop yield due to the unavailability of inputs, credit, high-yielding varieties, and irrigation water and increase in temperature. Gregory et al. [15] concluded that an increase of 1°C in temperature above 32°C can decrease yield of rice by 5%. These temperature effects were the most deleterious for the major crops like wheat, rice, and maize [16–18]. Additionally, increase in temperature also affects the wheat protein contents [19, 20].

The effect of climate change on productivity of crop shows the major role of agronomists to develop such varieties and cropping system that are more resilient to the climate change with high production. Modification in the crop due to change in climate was not significant, its might be due to gradually increase in carbon di oxide and temperature rate that modified the time of sowing, veracity and crop production management practices will allow some adaptation in the crop production system. These include various adaptations like the selection of crops that have strong mechanisms and high resistance against disease, are more resilient to the abiotic stresses like heat and temperature, and have stronger genetic enhancement to compete with changing environment and the selection of cropping system according to the current climatic condition. According to Tubiello et al. [21], increase in CO_2 concentration and temperature can decrease the yield of existing varieties by 10–40%. The combination of early planting of summer and spring crops can sustain the present yield of the crops [21]. Change in climate may cause change in water regimes, which may increase water demand in temperate regions while in tropical and subtropical regions, this may lead to water scarcity [22]. Further studies are needed to discover the most adoptive form of cropping system for specific regions keeping in view the climate change scenarios and for that agronomists need to work closely in the water management department. Change in climate may bring new disease, pests, and weeds that may cause serious problems for the crops. Some of the pests and weeds which are under economic injury level become problem by exploiting the changing condition [23]. Again, agronomists will need to work with the help of integrated pest management and integrated weed management and other tactics to help control the problematic weeds, pests, and diseases.

The major role of agronomy is discovering new techniques for higher crop production without depleting natural resources and intensifying climate change. Choices for enhancing crop production safely involved extensification and intensification [13]. Extensification will help to raise the total quantity of production and contribute to increases in production, but increase through extensification is limited due the availability of limited new land [15].

According to Greenland et al. [24], more than 3 billion hectares of land is available for cultivation and can be used for good production and about 1.2–1.5 billion hectares of land is already cultivated. In general, further agriculture extensification will cause very limited increase in crop production. Typically, further extensification will contribute just 7.4% to cereal production while estimated extensification to crop production ranges from 18% in South Asia to 47% in sub-Saharan Africa by the year of 2020. To decrease the intensification of climate change and increase the extensification and intensification of farming practices in the subtropical and tropical areas should need to change with more resilience ones [15].

5. Agronomy and food security

To assess the impacts of climate change on crop productivity and food security, agronomic research has thus provided an admirable basis. According to global harvest initiatives (2010), the global agricultural productivity must be increased by 1.75% to double the agricultural productivity by 2050. The average annual TFP growth rate in low-income countries is in trouble. However, Sustainable Development Goal 2 (SDG-2) calls for doubling of crop productivity for small-scale farmers in the lowest-income countries. But the current annual rate of TFP growth in low-income countries is just 0.96%. If this decline sustains for longer period, people in low-income countries will increase the use of soil and water, which are already threatened by extreme weather and climate change (**Figures 3** and **4**). Implementation of some farming practices has a significant impact and ecofriendly consequences at the watershed level. For example, growing of fruit trees on contours or other non-timber trees for the compensation of decline in crop yield could have a significant saving effect on water conservation and water use efficiency. This will provide a new way of farming to the farmers for increasing their income and will help to stabilize their socioeconomic status. Agronomists are needed to design such a productive agricultural system that is more suitable and resilient to the socio-economic needs of the farming communities like poverty and hunger alleviation, food security, climate change adaptation, and environmental protection.

Increase in crop productivity is important to reduce food security problem; therefore, agronomic research is related to food security. Food security can be ensured by an efficient system of food, food production, and new research in the area of crop production. Food system is a set of continuous interaction between and within humans and their biogeophysical environment and it includes food production, processing, and food allocation and food consumption [25], while agronomy has an important role in these activities like producing food, by modern scientific methods and practices, storage and processing of staple food, and production timing in relation to market and food diversity in terms of nutritional balance. To this end, agronomic research needs to be better linked to wide-ranging interdisciplinary sectors and across sectors of the food industry. This will facilitate the building of integrated socioeconomic-biophysical models that will enable analysis of adaptation

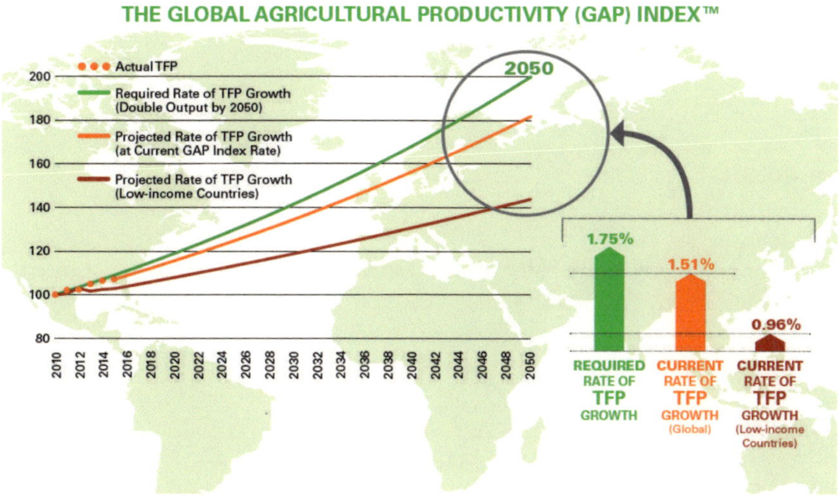

Figure 3.
Agricultural output from TFP growth.

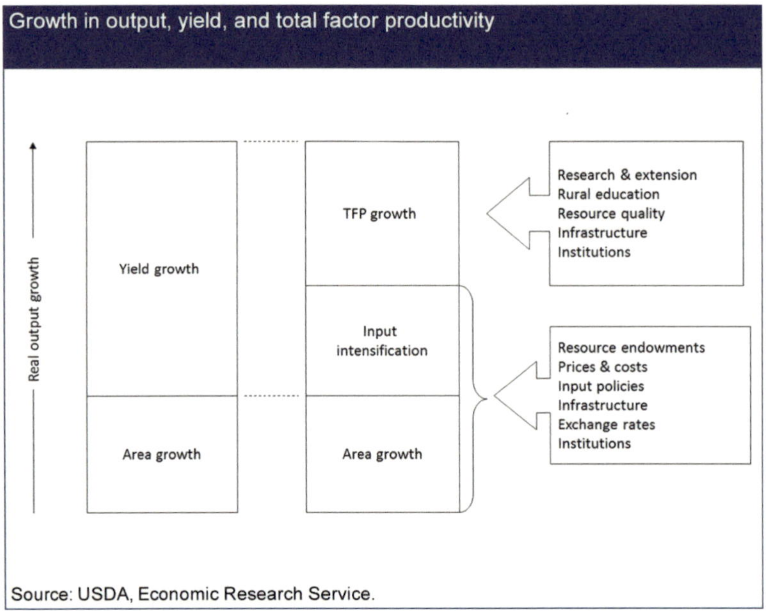

Figure 4.
Economic survey of USDA.

options to food systems, thereby underpinning policy formulation for improved food security and nutrition. The SDGs emphasize the importance of agriculture and the need to reinvigorate farming worldwide by supporting farmers, increasing investments in research, technology and market infrastructure, and extending knowledge sharing. This will catalyze innovation and empower farmers.

Author details

Amanullah and Shah Khalid
Department of Agronomy, Faculty of Crop Production Sciences, The University of Agriculture Peshawar, Pakistan

*Address all correspondence to: amanullah@aup.edu.pk

References

[1] Vitousek PM, Mooney HA, Lubchenco J, Melillo J. Human domination of earth's ecosystems. Science. 1997;**277**:494-499

[2] Steffen W, Tyson P. Global Change and the Earth System: A Planet under Pressure. Stockholm: International Geosphere—Biosphere Programme (IGBP); 2004

[3] Fischer G, Shah M, van Velthuizen H, Nachtergaele FO. Global Agro-Ecological Assessment for Agriculture in the 21st Century. Laxenburg, Austria: International Institute for Applied Systems Analysis; 2001

[4] Rosegrant MW, Cline SA. Global food security: Challenges and policy. Science. 2003;**302**:1917-1919

[5] Parry M, Rosenzweig C, Inglesias A, Livermore M, Gischer G. Effects of climate change on global food production under SRES emissions and socio-economic scenarios. Global Environmental Change. 2004;**14**:53-67

[6] Stern N. Review on the Economics of Climate Change. London: HM Treasury; 2006

[7] Hadley-Centre. Effects of Climate Change in the Developing Countries. UK: Met Office; 2006

[8] IPCC. Africa. In: McCarthy JJ, Canziani O, Leary NA, Dokken DJ, White KS, editors. Climate Change 2001: Impacts, Adaptation and Vulnerability. Contribution of Working Group II to the Third. 2001

[9] Tyson P, Odada E, Schulze R, Vogel CI. Regional–global change linkages: Southern Africa. In: Tyson P, Fuchs R, Fu C, Lebel L, Mitra AP, Odada E, Perry J, Steffen W, Virji H, editors. Global—Regional Linkages in the Earth System. Berlin, London: START/IHDP/IGBP/WCRP. Springer; 2002

[10] Doos BR. The problem of predicting global food production. Ambio. 2002;**31**:417-424

[11] UN-DESA. World population prospects: The 2004 revision report, chapter 1. United Nations Department of Economic and Social Affairs/ Population Division; 2004

[12] Tilman D, Cassman KG, Matson PA, Naylor R, Polasky S. Agricultural sustainability and intensive production practices. Nature. 2002;**418**:671-677

[13] Bruinsma J. World Agriculture: Towards 2015/2030. An FAO Perspective. London: Earthscan Publications Ltd; 2003

[14] Evans LT. Feeding the Ten Billion. Cambridge: Cambridge University Press; 1998

[15] Gregory PJ, Ingram JSI, Andersson R, Betts RA, Vrovkin V, Chase TN, et al. Environmental consequences of alternative practices for intensifying crop production. Agriculture, Ecosystems and Environment. 2002;**88**:279-290

[16] Jones PG, Thornton PK. The potential impacts of climate change on maize production in Africa and Latin America in 2055. Global Environmental Change. 2003;**13**:51-59

[17] Stige LC, Stave J, Chan KS, Ciannelli L, Pettorelli N, Glantz M, et al. The effect of climate variation on agro-pastoral production in Africa. Proceedings of the National Academy of Sciences. 2006;**103**:3049-3053

[18] Kurukulasuriya PR, Mendelsohn R, Hassan R, Benhin J, Deressa T, Diop M,

et al. Will Africa survive climate change? World Bank Economic Review; 2006

[19] Kettlewell PS, Sothern RB, Loukkari WL. UK wheat quality and economic value are dependant on the North Atlantic oscillation. Journal of Cereal Science. 1999;**29**:205-209

[20] Slingo JA, Challinor AJ, Hoskins GJ, Wheeler TR. Food crops in a changing climate. Philosophical Transactions of the Royal Society B. 2005;**360**:1983-1989

[21] Tubiello FN, Donatelli M, Rosenzweig C, Stockle CO. Effects of climate change and elevated CO_2 on cropping systems: Model predictions at two Italian locations. European Journal of Agronomy. 2000;**13**:179-189

[22] UNDP. The 2006 Human Development Report—Beyond Scarcity: Power, Poverty and the Global Water Crisis. New York. USA: United Nations Development Programme; 2006

[23] Sutherst R, Baker RHA, Coakley SM, Harrington R, Kriticos DJ, Scherm H. Pests under global change—Meeting your future landlords? In: Candell JG, editor. Terrestrial Ecosystems in a Changing World. Springer; 2007. p. 336

[24] Greenland DG, Gregory PJ, Nye PH. Land resources and constraints to crop production. In: Waterlow JC, Armstrong DG, Fowden L, Riley R, editors. Feeding a World Population of More than Eight. 1998

[25] Ericksen PJ. Conceptualizing food systems for global environmental change research. Global Environmental Change. 2008;**18**:234-245

Chapter 2

Factors Affecting Yield of Crops

Tandzi Ngoune Liliane and Mutengwa Shelton Charles

Abstract

A good understanding of dynamics involved in food production is critical for the improvement of food security. It has been demonstrated that an increase in crop yields significantly reduces poverty. Yield, the mass of harvest crop product in a specific area, is influenced by several factors. These factors are grouped in three basic categories known as technological (agricultural practices, managerial decision, etc.), biological (diseases, insects, pests, weeds) and environmental (climatic condition, soil fertility, topography, water quality, etc.). These factors account for yield differences from one region to another worldwide. The current chapter will discuss each of these three basic factors as well as providing some recommendations for overcoming them. In addition, it will provide the importance of climate-smart agriculture in the increase of crop yields while facilitating the achievement of crop production in safe environment. This goes in line with the second goal of 2030 Agenda for Sustainable Development of United Nations in transforming our world formulated as end hunger, achieve food security, improve nutrition and promote sustainable agriculture.

Keywords: crop, yield, production, food, agriculture, environment

1. Introduction

Agriculture is a key activity of human being since it provides basic needs such as food, clothing and shelter. It has been demonstrated that every 1% increase in agricultural yield translates into a 0.6–1.2% decrease in the numbers of absolute poor households in the world [1]. Meanwhile, population growth was predicted to be 9.7 billion by 2050 and this will require an increase of about 70% in food production to meet the demand [2]. Rainfed agriculture is projected to produce one-third or more of the food increase in global food output for the coming decades. Unfortunately, agricultural productivity depends on increasingly extreme weather phenomena. Thus, water availability, air pollution, and temperature have a large impact in agriculture [3].

Several factors pose significant risk to farms leading to yield reduction when they are not correctly monitored and well managed. These factors can be grouped into three categories which are technological, biological and environmental [4]. The pressure to increase crop production in many countries, has resulted in the expansion of land area dedicated to agriculture and the intensification of cropland management through practices such as irrigation, use of large quantities of inputs like inorganic fertilizers and synthetic chemicals for pest and weed control [5]. These practices have resulted in degradation of soil properties and water quality, acceleration of soil erosion, contamination of groundwater and decline of food quality. This has prompted sustainable intensification initiatives to increase yields on existing farmland while decreasing the environmental impact of agriculture [6–10].

Organic crop production is one of the alternative agricultural practices promoted for the reduction of environmental pollution. As a result, several countries have introduced organic farming practices to replace the chemical-dependent ones [11]. To conserve and regenerate soil properties, the maintenance of soil organic matter (SOM) has received considerable attention. Although SOM is considered key to soil health, its relationship with yield is contested because of local-scale differences in soils, climate, and farming systems. The relationship between these factors should be quantified and proper soil management strategies set up to ensure sustainable crop production [5].

The impact of climate change in our agricultural systems is undoubtable. For example, drought followed by intense rain can increase the flooding potential, thereby creating conditions that favor fungal infestations of leaves, roots and tuber crops. In addition, reduction of bees' density due to global climate change has led to local extinction of several plant species [12]. The production of enough food to match population growth while preserving the environment is a key challenge, especially in the face of climate change. This chapter will review factors affecting yields of crops and provide some strategies to overcome yield loss while preserving the environment.

2. Environmental factors affecting crop yields

The environmental factors affecting crop yields can be classified into abiotic and biotic constraints. Actually, these factors are more intensified with global warming which leads to climate change. Abiotic stresses adversely affect growth, productivity and trigger a series of morphological, physiological, biochemical and molecular changes in plants. The abiotic constraints include soil properties (soil components, pH, physicochemical and biological properties), and climatic stresses (drought, cold, flood, heat stress, etc.). On the other hand, biotic factors include beneficial organisms (pollinators, decomposers and natural enemies), pests (arthropods, pathogens, weeds, vertebrate pests) and anthropogenic evolution.

2.1 Abiotic constraints

2.1.1 Effects of climatic conditions on crops

Variations in annual rainfall, average temperature, global increase of atmospheric CO2, and fluctuations in sea levels are some of the major manifestations of climate change, which negatively impact crop yields [13]. Temperature and rainfall changes are expected to significantly have negative impact on wide range of agricultural activities for the next few decades. With the changing of climate, agriculture faces increasing problems with extreme weather events leading to considerable yield losses of crops. Most often, crop plants are sensitive to stresses since they were mostly selected for high yield, and not for stress tolerance. Climate change is the result of global warming. It has devastating effects on plant growth and crop yield which can affects directly, indirectly, and socio-economically reduce crop yields by up to 70% [14] (**Figure 1a**). Weather variations present positive and negative effects in the environment with very high expression of negative effects (**Figure 1b**).

The regression analysis model between historical climatic data and yield data for food crops over the last 30 years in Nepal showed an increase in temperature of approximately 0.02–0.07°C per year in different seasons and a mixed trend in precipitation [15]. Additionally, no significant impact of climate variables on yields of all crops was observed and the regression analysis revealed negative relationships

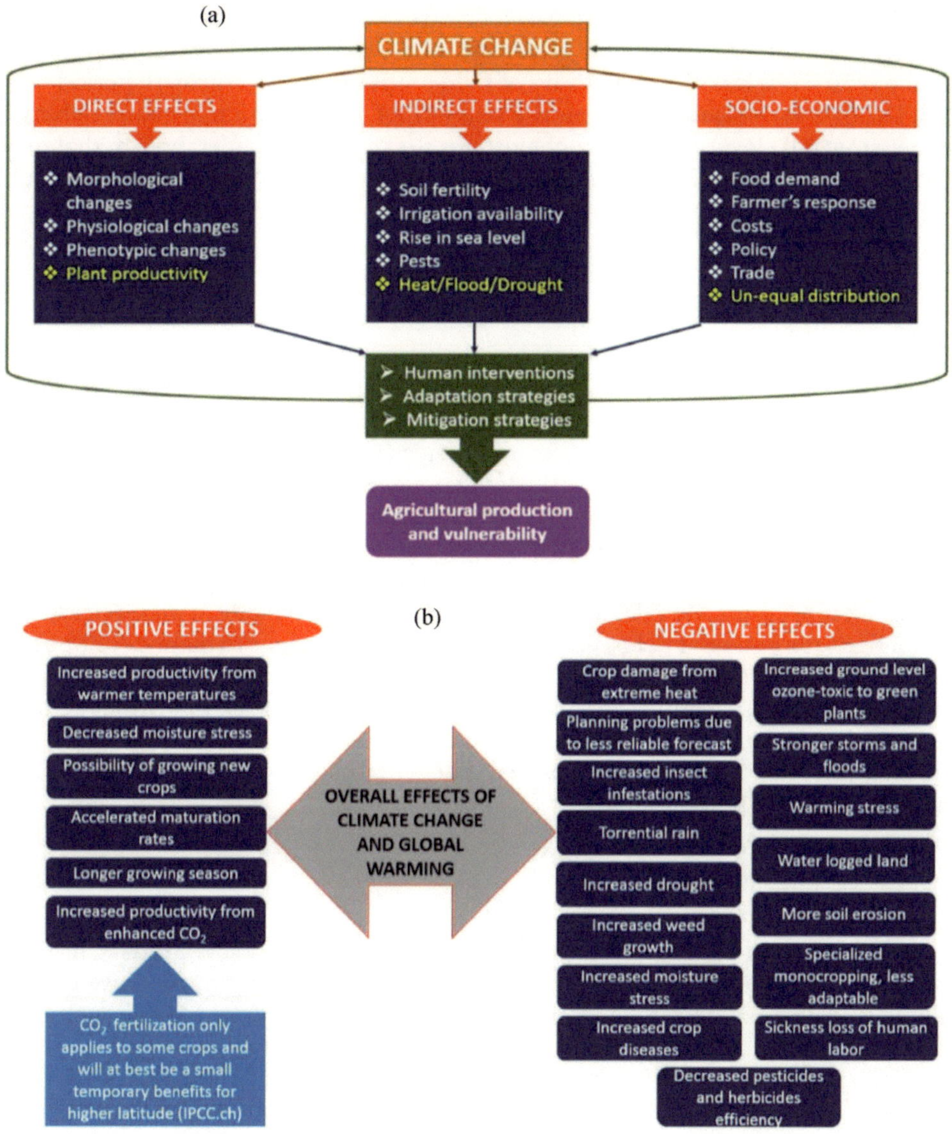

Figure 1.
General effects of climate change in agricultural production (a), the positive and negative impacts in the environment (b) [13].

between maize yield and summer precipitation, between wheat yield and winter minimum temperature, and finally positive relationship was observed between millet yield and summer maximum temperature.

2.1.1.1 Drought

Drought refers to a situation in which the amount of available water through rainfall and/or irrigation is insufficient to meet the evapotranspiration needs of the crop [16]. Climate change is driven by changes in water availability (volumes and seasonal distribution), and in water demand for agriculture and other competing sectors. The impending climate change adversities are known to alter the abiotic stresses like variable temperature regimes and their associated impacts on water availability leading to drought, increased diseases and pest's incidence and extreme weather events at local to regional scale [16]. Moisture or drought stress accounts for about 30–70% loss of productivity of field crops during crop growth period [16]. Drought stress can induce abscisic acid (ABA) accumulation in guard cells to trigger stomatal closure [17]. Drought also results in abnormal metabolism that may reduce plant growth, and/or cause the death of entire plant. Drought has different effects at different stages of plant growth with the most sensitive growth stage being flowering period.

2.1.1.2 Heat stress

Heat stress is the rise in temperature beyond a threshold level for a period sufficient to cause permanent damage to plant growth and development [18]. The Intergovernmental Panel on Climate Change (IPCC) projected rise of the temperature by 3–4° by 2050 [19, 20]. High temperature regimes due to climate change affect the percentage of seed germination, photosynthetic efficiency, crop phenology, reproductive biology, flowering times, pollen viability and pollinator populations [16, 21]. Under heat stress at reproductive growth stage, the increase of temperature prevents the swelling of pollen grains, which results in poor release of pollen from the anther at dehiscence. Heat stress is deleterious to plant developmental stages, including generation and function of reproductive organs. Furthermore, variable temperature regimes may result in unpredictable disease epidemics across geographic regions in the world. Heat stress contributed about 40% to overall yield loss of wheat [22], 1.0–1.7% yield loss per day in maize for every raise in temperature above 30°C [23].

2.1.1.3 Cold stress

Cold or chilling stress experiences by plants from 0 to 15°C [24], leads to major crop losses. Various types of crops in tropical or subtropical origin are injured or killed by non-freezing low temperatures, and exhibit different symptoms such as poor germination, stunted seedlings, chlorosis, or growth retardation, reduced leaf expansion and wilting and necrosis. In general, plants respond with changes in their pattern of gene expression and protein synthesis when exposed to low temperatures [25]. In general, plants from temperate climatic regions are considered to be chilling tolerant with variable degree compare to tropical and sub-tropical crops, and can increase their freezing tolerance by cold acclimation [26].

2.1.1.4 Soil properties

Soils are the uppermost part of the earth's crust, formed mainly by the weathering of rocks, formation of humus and material transfer. They vary in terms of origin, appearance, characteristics and production capacity. Soil fertility is the ability of a soil to deliver nutrients needed for the optimum growth of a specified crop. Soil fertility is one of the most important factors in crop production [10]. It has the ability to support crop production determined by the entire spectrum of its physical, chemical and biological attributes. Soil fertility is one important aspect of soil productivity since it is a major source of micronutrients (Fe, B, Cl, Mn, Zn, Cu, Mo, Ni) and macronutrients (N, P, K, Ca, S, Mg, C, O, H) that are needed for plant growth. The lack of these nutrients in the soil causes deficiencies in plants, and their excess leads to toxicities, which have negative impacts on crop yields.

Several parameters can be used to determine the fertility status of a soil. Among them, the soil fertility index was found to be the most useful indicator that helps to improve sustainable land use management and achieve economical yield in crop production [27]. In several regions in the world, some croplands have undergone human-induced soil degradation resulting in poor yield production per unit area of crop harvest. Around 40% of agricultural lands are affected by human induced land degradation. Intensive agricultural production characterized by overuse of fertilizers and chemicals without adherence to agricultural sustainability leads to a decline of soil health, land degradation and severe environmental problems [28]. It is important to note that the deterioration of soil fertility normally takes pace over several years.

2.1.1.5 Soil salinity and acidity stress

Salinity stress affects crop production in over 30% of irrigated crops and 7% of dry land agriculture worldwide [29]. It is one of the major problems affecting crop production all over the world since around 20% of cultivated land and 33% of irrigated land are salt-affected in the world [30]. Salt causes osmotic stress and ionic toxicity in crop plants. Under normal conditions, the higher osmotic pressure in plant cells permits the absorption of water and essential nutrients from a soil solution into the root cells. However, under salt stress conditions, the high concentration of salts in the soil solution prevents absorption of water and essential minerals but will facilitate the entry of Na^+ and Cl^- ions into the cells, which will have direct toxic effects on cell membranes as well as on metabolic activities in the cytosol [31].

Low soil pH increases as a result of release of acidifying aluminum, iron and manganese ions, leaching of base ions such as calcium, magnesium, potassium and sodium, decomposition of soil organic matter and regeneration of organic acids, nitrification of ammonia-based fertilizers [32, 33] as well as land management practices. Low soil pH significantly affects crop growth and therefore decreases yield. In maize for instance, soil acidity causes yield loss of up to 69% [34].

2.1.1.6 Floods

Floods entail different stressful conditions to plants, mainly depending on water depth and its duration. Soil waterlogging damages most crops, with the exception of rice, which like other wetland species thrives when plants are not completely submerged. In view of the changing climate, flooding has become frequent in many lowlands and cultivated areas every year and causes a lot of damage to human beings including losses in crop yields and food stuffs.

Flooding usually occurs with heavy rainfall, poor soil drainage and poor irrigation practices. Soil waterlogging has negative impacts on crop production especially for dryland species (such as most cereals, legumes, tubers, etc.) which include several crops. The excess water results in complex changes in plant physiology for non-adapted crops. This leads to restriction of gas diffusion between the plant and its surroundings (accumulation of high CO_2 and ethylene in the root zone with very low O_2), hypoxia (oxygen levels limit mitochondrial respiration) and anoxia (respiration is completely inhibited), often accompanied by increased of mobilization of 'phytotoxins' in reduced soils, leading to poor root metabolism (inability to absorb nutrients), lack of energy within plant cells, restriction of photosynthetic activities and therefore poor growth or death of plant roots and shoots.

The first constraint for plant growth under flooding conditions is the immediate lack of oxygen necessary to sustain aerobic respiration of submerged tissues [35–37].

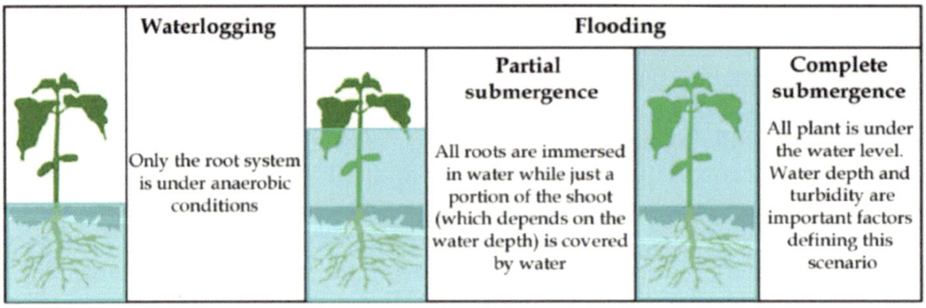

Figure 2.
Different levels of excess of water in crop environment [38].

As the duration of flooding increases, there is progressive decrease in soil reduction-oxidation potential (redox potential) [38] (**Figure 2**). Flooding events can be classified by two categories: waterlogging where only the root system inside the soil is affected [39]; and submergence, where also parts or the whole shoot are under water [40]. In tree species with different flooding sensitivity, the importance of root-to-shoot transport of metabolites to 'use rather than lose' is a relevant criterion used to identify the tolerant species [41]. Only non-wetland plants can survive flooding for a short period of time. The two survival strategies to flooding are plant avoidance of oxygen deficiency in tissues and the adaptation to oxygen deficiency [42].

3. Biotic factors affecting crop yields

3.1 Diseases and pests

Plant diseases are caused by different micro-organisms such as viruses, bacteria and fungi. In addition, various soil-borne and above ground insect pests also affect crop production. Variation in climatic conditions often favors the multiplication of pathogens while negatively affecting plant productivity and soil fertility. It causes the reduction of available resources for plants, which fail to produce enough biomass, seeds, and thus yield. Climate-driven migration allows the movement of pathogens and pests from one region to another. Thus, the locally adapted crop genotypes confront new biotic stress factors. The interaction of plants with microbes or microbe-associated molecular patterns can induce resistance to secondary infections by pathogens. This involves the production and systemic signal of a complex of low-molecular-weight plant metabolites, which are well described for dicotyledonous plants, but poorly understood for monocotyledonous plants such as cereal crops [43]. Because of climate variability and change, it is anticipated that new diseases and pests might appear, or that the virulence of the current types may increase.

The changing of the climate is bringing new types of diseases and pests that do not have any control methods yet. For example, maize lethal necrosis (MLN) is one of the most devastating diseases found in maize in Eastern and Central African countries. It is caused by the synergistic interaction between *Sugarcane Mosaic Virus* (SCMV) and *Maize Chlorotic Mottle Virus* (MCMV). It causes yield reduction ranging from 30 to 100% in farmers' fields depending on the time of infestation [44]. MLN is transmitted by beetles, rootworms, thrips, stem borers, several species of aphids in non-persistent manner, infected soil, infected seeds and any tools or materials used in the infected field [45]. Moreover, Russian Wheat Aphid (RWA) is one of the world's invasive pests of wheat, barley and other cereal grains. It is widespread in cereal growing regions of Africa, Asia, Europe, Middle East, North and South America, recently in Australasia [46]. The visual symptoms of infestation in plants are chlorosis, necrosis, wilting, stunting, leaf streaking with whitish, yellow and purple longitudinal leaf markings, trapped awns, rolled leaves and heads that fail to flower [46]. These pests have high resistance to extreme weathers events. RWA caused yield losses up to 80% in wheat and 100% in barley. The main challenge associated with the RWA is that new biotopes that are tolerant to available insecticides continue to appear. Some of the biotopes also overcome resistance of some crop varieties. Elevated atmospheric carbon dioxide has also been found to alter the efficacy of some biotopes. They are therefore constant threat to crop production.

4. Technological factors affecting crop yield

A wide range of technological innovations in agriculture like genetic improvement of varieties, fertilizer technology, adaptive microbial technology, pesticides, farm machinery, agronomic and management practices (integrated management of nutrients and pests) have been achieved through research programs to understand their implications in enhancing crop productivity [16]. It has been reported that 1 kg of nutrient fertilizer produces 8 kg of grain [47]. In addition, fertilizers are commonly believed to be very important in crop production since they contribute up to 50% of the crop harvest product [48]. The doubled increase of food production worldwide was partially attributed to a 6.9-fold increase in nitrogen fertilization and a 3.5-fold increase in phosphorous fertilization in the 1990s [49].

Different factors have negative influence in agricultural practices. In Bangladesh, farmers were given chemical fertilizers and pesticides at a subsidized price and therefore increased fertilizer application to enhance crop yield. In the Philippines, because of the huge amount of lime and urea used by farmers over years, the sugarcane farms developed lime layer in the subsoil, which caused phosphorous deficiency while banana farms have excessive potash, which created an imbalanced ratio of potassium and magnesium. The average yield production of sesame in Jigawa State was reported to be 0.6 t/ha instead of 1.25 t/ha under well-managed farms [50]. In general, the application of inappropriate agronomic practices such as untimely planting, incorrect plant spacing, wrong method of planting, poor sowing depth, delayed weeding, ineffective pest and disease control, inappropriate use of fertilizers, untimely harvesting and use of low yielding varieties, will always significantly reduce crop yields.

5. Strategies to overcome crop yield reduction

Climate smart agriculture (CSA) is now widely accepted as the best approach for addressing the effects of climate change in agriculture. It is defined as agriculture that sustainably increases productivity, resilience (adaptation), reduces/removes greenhouse gases (mitigation), and enhances the achievement of national food security and development goals. CSA promotes the transformation of agricultural systems and requires the transformation of agricultural policies to increase food production, to enhance food security, to ensure that food is affordable (low input-cost) while ensuring sustainable natural resource management and resilience to a changing climate.

5.1 Management of the environment

Climate influences all components of crop production including crop area and crop intensity. Weather forecasting and crop yield prediction or simulations are relevant tools that provide a warning to farmers in preparation of the upcoming season. From the simulation results, farmers can change the crop planting date, use appropriate genotypes, adjust the fertilization and the irrigation cycles to obtain reasonable yields, thus reducing the risk of unexpected events [51]. Several studies have been successfully conducted in crop yield simulation models and were reviewed by Tandzi and Mutengwa [51]. In a general view, the reduction of chemicals' usage such as fertilizers and pesticides, associated with the improvement of crop input use efficiency will minimize greenhouse gases emissions while protecting the environment. It has been reported that any programs that are working to minimize the adverse impact of climate change on food crops production should first consider the type of crop grown, the production area as well as the geographical and climatic conditions [15]. The knowledge of appropriate planting methods is important because climate events influence the selection of planting method and thus yield even though the total planted area remains unchanged [52]. There is a possibility of producing more yields in sustainable agriculture while generating less environmental pressure (**Figure 3**).

5.2 Management of agricultural inputs

Improvement of irrigation performance and water management are critical to ensure the availability of water both for food production and for competing human and environmental needs. To improve crop productivity and sustainability, it is very important to evaluate the effects of human activities in soil fertility through the use of appropriate agricultural systems such as tillage, use of recommended rates and types of fertilizer, incorporation of farmyard manure and/or crop residues into the soil (increase supply of N, P, K and other nutrients) and avoid sewage sludge irrigation. The application of these inputs improves physical properties of soil or soil organic matter in the long term and ensures sustainable agriculture. Shang et al. [28] found that high crop yields and low production variability can be achieved by increasing integrated soil fertility quality index in intensive cropping systems.

Climate-smart agriculture is the best way to lower the negative impact of climatic variations on crop adaptation. The type of inputs utilized during production combined with adapted high-yielding genotypes will determine the quality and quantity of harvest products to obtain (**Figure 4**). In addition, cover crops provide weed and pathogen control, decreased soil erosion, reduced loss of soil nitrogen, phosphorus and carbon. On the other hand, plant-beneficial microbes provided disease control and phosphorus availability [53]. The application of integrated pests and diseases management in farmers' fields will consistently reduce yield loss. Alternative agricultural practices such as organic production is promoted as being environmentally friendly with reduce agricultural impacts on water quality. Several countries have introduced organic farming practices to produce good quality food. The application of compost with chemical fertilizers not only results in high yields but also improves soil organic matter accumulation and soil fertility. In addition, the application of chicken manure compost enhanced soil quality and increased the accumulation of soil organic matter, available phosphorus (P), potassium (K), calcium (Ca), and magnesium (Mg) content in Botswana [54]. Microbial fertilizers are distinctly environment-friendly, non-bulky, cost-effective and play a significant role in plant nutrition [55]. Policymakers in different countries should formulate policy on sustainable fertilizer and pesticide management in crop production with

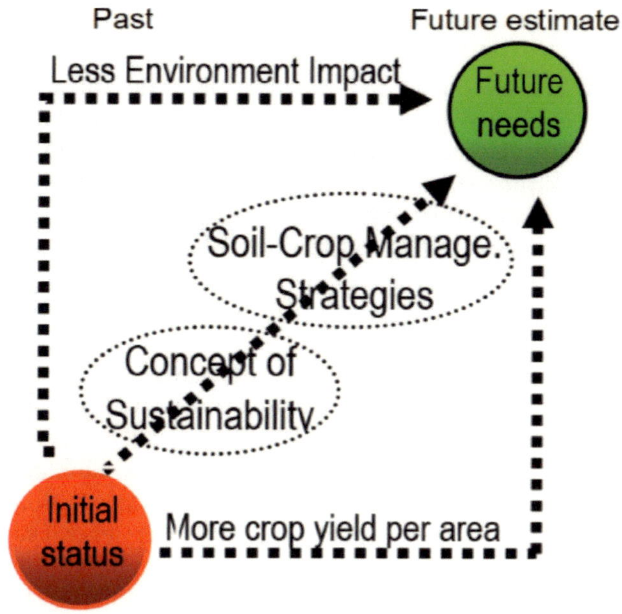

Figure 3.
Strategy of moving towards higher crop productivity and less environmental impacts [28].

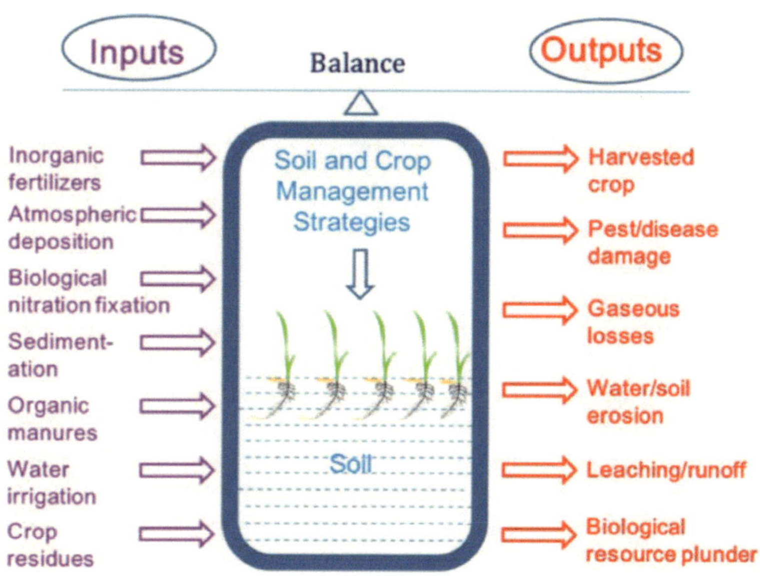

Figure 4.
Nutrient budgets between inputs and outputs [28, 56, 57].

different placement methods to reduce the overuse of those chemicals while preserving the environment. Guan et al. [58] identified RCF3, a KH domain-containing RNA-binding protein localized in the nucleus, as an upstream negative regulator of thermo-tolerance by modulating the expression of genes encoding heat-shock proteins (HSPs) in Arabidopsis. In South Africa, the maintenance of yield quality and quantity under actual prevailing environmental conditions have been largely achieved through the change in water and fertilizer management as well as new crop management practices (such as appropriate use of rotation system, lower seeding and fertilizer application).

5.3 Development of new adapted crop genotypes

Breeding is routinely conducted to increase levels of durable resistance to specific pests, diseases and different abiotic stresses using conventional crop improvement methods. However, there is now an increased use of modern biotechnology techniques such as marker-assisted selection, and transgenic approaches that involve genetic modification and high-throughput sequencing of both plant and pathogenic micro- organisms. Attempts have also been made to utilize transgenic technologies to build intrinsic tolerance mechanisms by the plants through alteration of functional genes [16]. Sustainable technologies like classical breeding approaches and integrated farming principles are also being considered to develop crops adaptation and/or enhance the adaptive mechanisms.

Under stress conditions, crop plants have evolved a set of perception and signal mechanisms to respond or adapt to adverse environmental conditions via regulation, transcription, gene expression, protein translation, modification, degradation, and metabolic regulation [17]. For example, strong associations were observed between the Na^+ content and some metabolites, including several sugars, suggesting that metabolic regulation is important for plant responses to salinity stress [59]. It has been demonstrated that manipulation of auxin biosynthesis pathway may improve crop plants tolerance to drought [60]. Physiological plant responses of crops to drought and heat stresses involve mechanisms to prevent membrane, regulate photosynthesis, respiration, and transpiration. For instance, developing crop genotypes with improved water used efficiency is one of the solutions to overcome drought stress. The most promising traits that might enhance crop flooding tolerance and facilitate longitudinal oxygen transport to sustain root aeration and water absorption in anaerobic soils, are anatomical adaptations such as formation of aerenchyma, a barrier against radial oxygen loss, and the growth of adventitious roots [39, 42]. The CBF/DREB1 genes are thought to be activators that integrate several components of the cold acclimation response by which plants increase their tolerance to low temperatures after exposure to non-freezing conditions. The DREB1/CBF genes have been successfully used to improve abiotic stress tolerance in a number of different crop plants [25].

The combination of genomics approaches such as marker-assisted selection (MAS) and genome wide associated studies (GWAS) can be efficiently used to develop biotic and abiotic stress tolerant cultivars (**Figure 5**). Future bio-computational integration of multiple omics and meta-omics with innovative research tools (reference genomes, proteomes, metabolomes with comprehensive annotations and structure–function relationships) will improve the understanding of the complexity of plant stress physiology [43] which will gather the development of the high-yielding and most adapted crop cultivars.

In definitive, there is a need to improve research activities into water quality and water use efficiency, nutrient and soil conservation technologies and techniques, climate-resistant crops and livestock, as well as agricultural productivity in line with the national development policy of each country, to promote the development of climate-smart agriculture which lower agricultural emissions and boosts agricultural production.

5.4 Climate: smart agriculture and food security

One of the most difficult and important tasks is to ensure the protection of the planet from the degradation through sustainable consumption and production, sustainable management of natural resources and urgent action to take towards climate change at national, regional and global level. Climate change is one of the leading

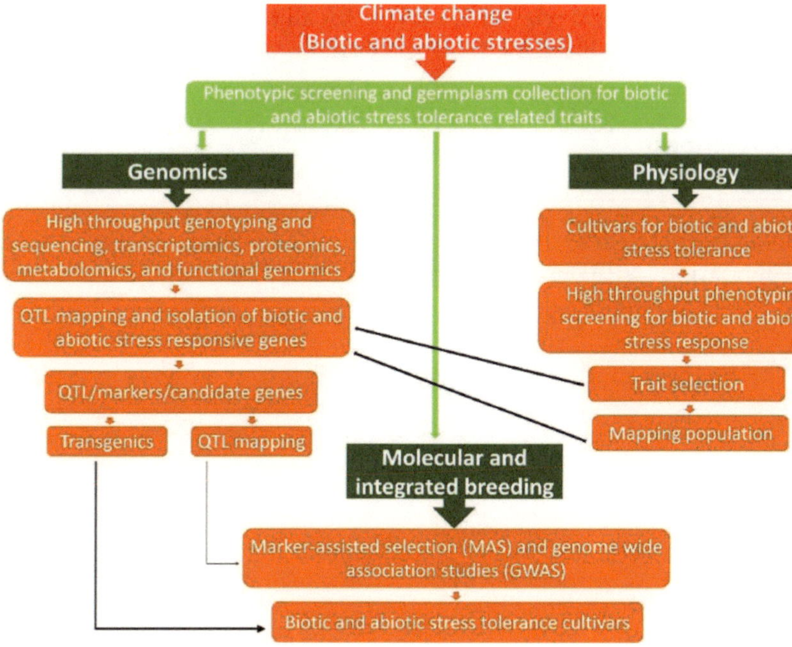

Figure 5.
Different steps of applying combined biotechnological tools in the breeding for biotic and abiotic stress tolerant crop genotype [13].

risks affecting the four dimensions of food security which are food availability, food accessibility, food utilization and food system stability [61]. Climate-smart agriculture (CSM) is an approach for transforming and reorienting agricultural systems to support food security under the new realities of climate change [62]. It promotes multidisciplinary actions to be taken by farmers, researchers, private sectors, civil society and policymakers towards climate-resilient pathways. In addition, CSM is based on three principles which are production (sustainable increase of the level of agricultural production and income), adaptation (development of resilient production systems adapted to climate change) and mitigation (reduction or elimination of greenhouse gas emission where possible) [63]. It is therefore a response to the challenges faced to satisfy the food needs of an increasing population in a changing climate.

6. Conclusion and recommendations

Climate smart agriculture sustainably increased crop yields while facilitating achievement of adaptation and mitigation goals in crop production. The development of new climate resilient crop tolerant and adapted to biotic and abiotic stresses will require the propagation of novel cultural methods, the implementation of various cropping schemes, and the combination of different conventional and non-conventional approaches. The development of integrated soil-crop system management and integrated diseases and pests' management with existing crop varieties and the increase of new improved and adapted high-yielding varieties under water and nutrient limited environment should be the new target for the coming generations. The application of genetically engineered crop plants by the introduction and/or overexpression of selected genes seem to be a viable option to hasten the breeding of improved adapted and high-yielding crop genotypes. Trans and interdisciplinary researches are needed to find relevant solutions for all the environmental challenges reducing crop yields while ensuring food security.

Acknowledgements

We appreciate the funding provided by the Govan Mbeki Research and Development Centre (GMRDC) at the University of Fort Hare and by the National Research Foundation (NRF) in South Africa.

Author details

Tandzi Ngoune Liliane[1,2]* and Mutengwa Shelton Charles[1]

1 Department of Agronomy, Faculty of Science and Agriculture, University of Fort Hare, Alice, South Africa

2 Institute of Agricultural Research for Development (IRAD), Yaounde, Cameroon

*Address all correspondence to: tnliliane@yahoo.fr

References

[1] Thirtle C, Irz X, Lin L, Mckenzie-Hill V, Wiggins S. Relationship between changes in agricultural productivity and the incidence of poverty in developing countries. In: DFID Report No. 7946. 2001

[2] Leung H. Stressed genomics—Bringing relief to rice fields. Curr. Opion. Plant Biol. 2008:201-208

[3] Noya I, González-García S, Bacenetti J, Fiala M, Moreira MT. Environmental impacts of the cultivation-phase associated with agricultural crops for feed production. Journal of Cleaner Production. 2018;**172**:3721-3733

[4] Metclfe DS, Elkins DM. Crop Production: Principles and Practices. New York: Macmillan Publishing Co., Inc.; 1980

[5] Oldfield EE, Bradford MA, Wood SA. Global meta-analysis of the relationship between soil, organic matter and crop yields. The Soil. 2019;**5**:15-32. DOI: 10.5194/soil-5-15-2019

[6] Allen P, Van Dusen D. Sustainable agriculture: Choosing the future. In: Appen P, Van Dusen D, editors. Global Perspectives on Agro-Ecology and Sustainable Agricultural Systems. Santa Cruz, CA: University of California; 1988. pp. 1-14

[7] Godfray HCJ, Beddington JR, Crute IR, Haddad L, Lawrence D, Muir JF, et al. Food security: The challenge of feeding 9 billion people. Science. 2010;**327**:812-818. DOI: 10.1126/science.1185383

[8] Foley JA, Ramankutty N, Brauman KA, Cassidy ES, Gerber JS, Johnston M, et al. Solutions for a cultivated planet. Nature. 2011;**478**: 337-342. DOI: 10.1038/nature10452

[9] Mueller ND, Gerber JS, Johnston M, Ray DK, Ramankutty N, Foley JA. Closing yield gaps through nutrient and water management. Nature. 2012;**490**:254-257. DOI: 10.1038/nature11420

[10] Wang C-H. Farming methods effects on the soil fertility and crop productionn under a rice—Vegetables cropping sequences. Journal of Plant Nutrition. 2014;**37**:1498-1513. DOI: 10.1080/01904167.2014.881876

[11] Oquist KA, Strock JS, Mulla DJ. Influence of alternative and conventional farming practices on subsurface drainage and water quality. Journal of Environmental Quality. 2007;**36**:1194-1204

[12] Ajetomobi J. Effects of weather extremes on crop yields in Nigeria. African Journal of Food, Agriculture, Nutrition and Development. 2016;**16**(4):11168-11184. DOI: 10.18697/a jfand.76.15685

[13] Raza A, Razzaq A, Mehmood SS, Zou X, Zhang X, Lv Y, et al. Impact of climate change on crop adaptation and strategies to tackle its outcome: A review. Plants. 2019;**8**(34):1-29. DOI: 10.3390/plants8020034

[14] Boyer JS. Plant productivity and environment. Science. 1982;**218**:443-448

[15] Poudel S, Shaw R. The relationships between climate variability and crop yield in a mountainous environment: A case study in Lamjung District, Nepal. Climate. 2016;**4**(13):1-19. DOI: 10.3390/cli4010013

[16] Kumaraswamy S, Shetty PK. Critical abiotic factors affecting implementation of technological innovations in rice and wheat production: A review. Agricultural Reviews. 2016;**37**(4): 268-278. DOI: 10.18805/ag.v37i4.6457

[17] Li W, Cui X. A special issue on plant stress biology: From model species to crops. Molecular Plant. 2014;7:755-757

[18] Hasanuzzaman M, Nahar K, Alam MM, Roychowdhury R, Fujita M. Physiological, biochemical and molecular mechanisms of heat stress tolerance in plants. International Journal of Molecular Sciences. 2013;**14**:9643-9684

[19] IPCC. Climate change and biodiversity. In: IPCC Technical Paper V. 2002. pp. 1-86

[20] Meeh GA, Stocker TF, Collins WD. Climate change 2007: The physical science basis. In: Fourth Assessment Report of the Intergovernmental Panel on Climate Change. Cambridge, UK: University Press; 2007

[21] Antle JM, McGuckin T. Technological innovations, agricultural productivity and environmental quality. In: Carlson GA, Zilberman D, Miranowski J, editors. Agric. Environ. Resour. Econ. 1993. pp. 175-220

[22] Elbasyoni IS. Performance and stability of commercial wheat cultivars under terminal heat stress. Agronomy. 2018;**8**:37

[23] Lobell DB, Banziger M, Magorokosho C, Vivek B. Nonlinear heat effects on African maize as evidenced by historical yield trials. Nature Climate Change. 2011;**1**:42-45

[24] Yadav SK. Old stress tolerance mechanisms in plants. A review. Agronomy for Sustainable Development. 2010;**30**:515-527. DOI: 10.1051/agro/2009050

[25] Sanghera GS, Wani SH, Hussain W, Singh NB. Engineering cold stress tolerance in crop planrs. Current Genomics. 2011;**12**:30-43

[26] Levitt J. Responses of plants to environmental stress. In: Chilling, Freezing, and High Temperature Stress. New York: Academic Press; 1980

[27] Shang Q, Ling N, Feng X, Yang X, Wu P, Zou J, et al. Soil fertility and its significance to crop productivity and sustainability in typical agroecosystem: A summary of lon-term fertilizer experiments in China. Plant and Soil. 2014;**381**:13-23. DOI: 10.1007/s11104-014-2089-6

[28] Shah F, Wu W. Soil and crop management strategies to ensure higher crop productivity within sustainable environments. Sustainability. 2019;**11**(1485):1-19. DOI: 10.3390/su11051485

[29] Schroeder JI, Delhaize E, frommer WB, Guerinot ML, Harrison MJ, Herrera-Estrella L, et al. Using membrane transporters to improve crops for sustainable food production. Nature. 2013;**497**:60-66

[30] Machado RMA, Serralheido RP. Soil salinity: Effect on vegetable crop growth. Management practices to prevent and mitigate soil salinization. Horticulturae. 2017;**3**(30):1-13. DOI: 10.3390/horticulturae3020030

[31] Kumar M. Crop plants and abiotic stresses. Biomolecular Research & Therapeutics. 2013;**3**(1):1000e125

[32] Heggeenstaller A. Managing soil pH for crop production. Crop Insight. 2012;**22**(15):1-4

[33] Johnson SL, Lincoln DE. Allocation responses in CO_2 enrichment and defoliation by a native annual plant *Heterothea subaxllaris*. Global Change Biology. 2000;**6**:767-778

[34] Tandzi NL, Mutengwa CS, Ngonkeu ELM, Gracen V. Breeding maize for tolerance to acidic soils: A review. Agronomy. 2018;**8**(84):2-21. DOI: 10.3390/agronomy8060084

[35] Armstrong W. Aeration in higher plants. Advances in Botanical Research. 1979;7:225-332

[36] Vartapetian BB, Jackson M. Plant adaptations to anaerobic stress. Annals of Botany. 1997;**79**:3-20

[37] Voesenek LACJ, Rijnders J, Peeters AJM, Van de Steeg HMV, De Kroon H. Plant hormones regulate fast shoot elongation under water: From genes to communities. Ecology. 2004;**85**:16-27

[38] Striker GG. Flooding stress on plants: Anatomical, morphological and physiological responses. In: Botany. 2012. pp. 1-27

[39] Bailey-Serres J, Colmer TD. Plant tolerance of flooding stress—Recent advances. Plant, Cell and Environment. 2014;**37**:2211-2215. DOI: 10.1111/pce.12420

[40] Sasidharan R, Bailey-Serres J, Ashikari M, Atwell BJ, Colmer TD, Fagerstedt K, et al. Community recommendations on terminology and procedures used in flooding and low oxygen stress research. The New Phytologist. 2017;**214**:1403-1407

[41] Kreuzwieser J, Rennenberg H. Molecular and physiological responses of trees to waterlogging stress. Plant, Cell & Environment. 2014;**37**:2245-2259

[42] Mustroph A. Improving flooding tolerance of crop plants. Agronomy. 2018;**8**(160):1-25. DOI: 10.3390/agronomy8090160.

[43] Dresselhaus T, Hückelhoven R. Biotic and abiotic responses in crop plants. Agronomy. 2017;**8**(267):1-6. DOI: 10.3390/agronomy8110267

[44] Karanja J, Derera J, Gubba A, Mugo S, Wangai A. Response of selected maize inbred germplasm to maize lethal necrosis disease and its causative viruses (sugarcane mosaic virus and maize chlorotic mottle virus) in Kenya. The Open Agriculture Journal. 2018;**12**:2015-2226. DOI: 10.2174/1874331501812010215

[45] Mekureyaw MF. Maize lethal necrosis desease: And emerging problem for maize production in eastern Africa. Journal of Plant Physiology & Pathology. 2017;**5**(4). DOI: 10.4172/2329-955X.1000170

[46] Umina P, Baker G, Edwards O. Russian Wheat Aphid: Tactics for Future Control. Grains Research and Development Corporation; 2017. 40 p. Project code: ACO00020-B

[47] Rehman A, Chandio AA, Hussain I, Jingdong L. Fertilizer consumption, water availability and credit distribution: Major factors affecting agricultural productivity in Pakistan. Journal of the Saudi Society of Agricultural Sciences. 2017;**18**:269-274

[48] Tomich TP, Kilby, Johnson B. Transforming Agrarian Economies: Opportunities Seized. Ithaca, NY: Cornell University Press; 1995. Opportunities missed

[49] Tilman D, Reich PB, Knops J, Wedin D, Mielke T, Lehman C. Diversity and productivity in a long-term grassland experiment. Science. 2001;**294**:843-845

[50] OLAM. Sesame production in Nigeria. In: Out Growers Training Manual for Jigawa State. 2007. pp. 3-6

[51] Tandzi NL, Mutengwa CS. Estimation of maize (*Zea mays* L.) yield per harvest area: Appropriate method. Agronomy MDPI. 2019

[52] Lizumi T, Ramankutty N. How do weather and climate influence cropping area and intensity? Global Food Security. 2014;**4**:46-50

[53] Robert DP, Matto AK. Sustainable agriculture enhancing environmental benefits, food nutritional quality and building crop resilience to abiotic and biotic stresses. Agriculture. 2018;**8**(8): 1-24. DOI: 10.3390/agriculture80100008

[54] Dikinya O, Mutwanzala N. Chicken manure-enhanced soil fertility and productivity: Effects of application rates. Journal of Soil Science and Environmental Management. 2014;**1**(3):46-54

[55] Aboudrare A. Agronomie Durable. In: Principes ET Pratiques. Rapport de Formation Continue. FAO; 2009. 49 p

[56] Gruhn P, Goletti F, Yudelman M. Integrated nutrient management, soil fertility and sustainable agriculture: Current issues and future challenges. In: IFRPI 2020 Vision Brief. Washington, DC, USA: IFRPI; 2000

[57] Wu W, Ma BL. Integrated nutrient management (INM) for sustaining crop productivity and reducing environmental impact: A review. Science of the Total Environment. 2015;**512**:415-427

[58] Guan Q, Lu X, Zeng H, Zhang Y, Zhu J. Heat stress induction of miE398 triggers a regulatory loop that is critical for thermotolerance in Arabidopsis. The Plant Journal. 2013;**74**(5). DOI: 10.1111/tpj.12169

[59] Hill cB, Jha D, Bacic A, Tester M, Roessner U. Characterization of ion contents and metabolic responses to salt stress of different Arabidopsis AtHKT1;1 genotype and their parental strains. Molecular Plant. 2013;**6**:350-368

[60] Kim J, Baek D, Park HC, Chun HJ, Oh D-H, Lee MK, et al. Overexpression of Arabidopsis YUCCA6 in potato results in high-auxin developmental phenotypes and enhanced resistance to water deficit. Molecular Plant. 2013;**6**:337-349

[61] FAO. Climate Change and Food Security: A Framework Document. Food and Agricultural Organization of the United Nations; 2008. 107 p. Available at: www.fao.org/forestry/15538-079b31d45081fe9c3dbc6ff34de4807e4.pdf

[62] Lipper L, Thornton PK, Campbell BM, Baedeker T, Braimoh A, Bwalya M, et al. Climate-smart agriculture for food security. Nature Climate Change. 2014;**4**:1068-1072. DOI: 10.1038/nclimate2437

[63] Muller C, Salgado R, Duran M, Le Coq JF, de Varax M, Gamba-Trimiño C, et al. Innovation platform for climate-smart agriculture in Honduras. In: CCAFS Policy Brief. Wageningen, the Netherlands: CGIAR Research Program on Climate Change, Agriculture and Food Security (CCAFS); 2018

Chapter 3

Possible Impacts of Climate Change on Sunflower Yield in Turkey

Hudaverdi Gurkan, Yasin Ozgen, Nilgun Bayraktar, Huseyin Bulut and Mustafa Yildiz

Abstract

Sunflower (*Helianthus annuus* L.) is the main raw material used to produce oil for consumption and oilseed in Turkey; however, its production is not sufficient, even for only domestic consumption. Therefore, studies were needed to determine how to increase both the production area and yield in Turkey. The aim of the study was to evaluate the possible effects of climate changes on future sunflower yield. A total of 29 provinces with intense sunflower cultivation during years of 1985–2014 were evaluated. Sunflower production values and meteorological data, which belong to years of 1985–2014, on climate projections, based on HadGEM2-ES Global Climate Model and RCP8.5 scenario that cover period of 2016–2099, were used as material. In the first part of the study, linear regression analyses were conducted between the observation and production data using the least squares method. In the second part, the possible effects of climate changes on sunflower yield for 2016–2040, 2041–2070, and 2071–2099 were determined using regression equations and climate projection data. Projections indicate that decreases in yield are expected, especially in the second half of this century. In Tekirdag and Konya provinces, where there is intensive sunflower cultivation, severe decreases in yield are expected for all studied periods.

Keywords: climate change, sunflower yield, regression analysis, HadGEM2-ES, RCP8.5

1. Introduction

Sunflower (*Helianthus annuus* L.) cultivation began after World War II after immigrants brought the plant to the Thrace region in Turkey. Thanks to the plant is drought resistant, sunflowers have a wide adaptability to climate changes and have become the main raw material used as a source of vegetable oil in Turkey.

According to TURKSTAT data, from 2004 to 2018, both cultivation area and yield in sunflower production have increased. During this period, the cultivated areas increased approximately by 32.4%, while the yield increased by 56.5%. The production volume also increased by 108.1% [1]. According to 2018 production data, sunflower (oil type) cultivation has produced 1.8 MT on 649 kha. The mean yield was 277 kg/Da; however, because of the increase in the population of Turkey and changes in dietary habits, sunflower production cannot meet domestic consumption. According to the product balance reports published by TURKSTAT, as of 2018, the domestic sufficiency ratio of sunflowers was 64.3% [2], which suggests that the oilseed production and vegetable oil sectors rely heavily on foreign sources.

Because of its ability to adapt to drought conditions, most of the sunflowers are cultivated under rainfed conditions. According to 2018 data, 64% of sunflower (oil type) is produced under rainfed agricultural conditions, whereas 36% are produced under irrigated agricultural conditions. Given that most production is based on rainfed farming conditions, the crop tends to be sensitive to changes in climate; therefore, serious differences have been observed in production volume and yield over several years. The 2007–2008 drought in Turkey caused significant losses in sunflower yield in addition to that in many other crops. Sunflower (oil type) production in 2007 decreased by 23.8% compared to that in the previous year.

It is inevitable that sunflower production, which is directly affected by climate change factors, will also be affected by the predicted climate changes. According to the climate change reports published by Turkish State Meteorological Service (TSMS), expected changes in rainfall and precipitation regime are predicted with increases in temperatures. For 2016–2099 and based on climate projections, the average temperature is expected to increase by 1.5–3.7°C. The total annual precipitation is expected to regionally decrease; however, in general, changes in precipitation balance during the year have been projected [3, 4].

Turkey is located within an area sensitive to climate change. Climate change in Turkey would inevitably affect the country's agriculture, and numerous studies have been conducted to determine these impacts [5–10].

The aim of this study was to determine the possible effects of climate change on sunflower (*Helianthus annuus* L.) yield, which would continue to have increasing importance in terms of Turkey's agriculture and economics.

2. Materials and methods

2.1 Material

2.1.1 Sunflower production

In the current study, TURKSTAT yield data on 29 provinces (Adana, Afyon, Aksaray, Amasya, Ankara, Aydın, Balıkesir, Bilecik, Bursa, Çanakkale, Çorum, Diyarbakır, Edirne, Eskişehir, İstanbul, İzmir, Karaman, Kayseri, Kırklareli, Kırşehir, Kocaeli, Konya, Kütahya, Osmaniye, Sakarya, Samsun, Tekirdağ, Tokat, and Uşak) in which there is intensive sunflower production were discussed (**Figure 1**).

To best determine the relationship between climate factors and yield, 1985–2014 (30 years) were considered and analyzed.

2.1.2 Meteorological parameters

The World Meteorological Organization (WMO) acknowledges that to evaluate the climate characteristics of a region, it is necessary to consider those characteristics for 30 years; therefore, 30-year meteorological parameters were preferred in determining their effects on plant yield.

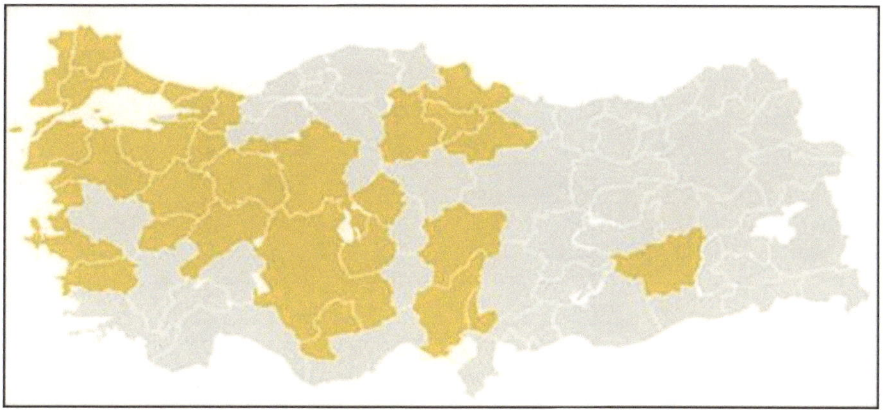

Figure 1.
Studied sunflower production areas.

In this study, the parameters that are supposed to affect yield were selected by considering the climate demands of the crop. The selected meteorological parameters that cover province-based sunflower vegetation period data were obtained from TSMS (2015). To determine the relationship between yield and climate factors, the following parameters were used:

- Number of days with daily minimum temperature ≤−5°C
- Monthly average temperature (°C)
- Number of days with daily maximum temperature >35°C
- Monthly average relative humidity (%)
- Number of days with daily average relative humidity >70%
- Monthly total sunshine duration (h)
- Monthly total precipitation (mm)

2.1.3 HadGEM2-ES global climate model

The most important tool in predicting the future climate is the modeling of climate [11]. Climate modeling studies have been conducted to determine the effects of climate changes that may occur in future periods. In Turkey, climate modeling studies were conducted within TSMS, and the final results were shared in 2015. In this study, the data on the selected meteorological parameters related to the 20-km-resolution climate projections were used on the basis of the report from HadGEM2-ES global model data and the RCP8.5 scenario used in "Turkey Climate Projections with New Scenario's and Climate Change (TR2015-CC)" by TSMS [3] for Turkey and the neighboring region. The RCP8.5 scenario is the scenario with the highest predictive radiation forcing and greenhouse gas concentration. In other words, RCP8.5 expresses the most pessimistic condition for the future periods. In this scenario, the radiative forcing reaches 8.5 w/m^2 in 2100 and continues to

Z matrix								W matrix	X matrix
$\sum s_i^2$	$\sum p_i s_i$	$\sum h_i s_i$	$\sum k_i s_i$	$\sum t_i s_i$	$\sum m_i s_i$	$\sum v_i s_i$	$\sum s_i$	$\sum s_{iy}$	A
$\sum s_i p_i$	$\sum p_i^2$	$\sum h_i p_i$	$\sum k_i p_i$	$\sum t_i p_i$	$\sum m_i p_i$	$\sum v_i p_i$	$\sum p_i$	$\sum p_i$	B
$\sum s_i h_i$	$\sum p_i h_i$	$\sum h_i^2$	$\sum k_i h_i$	$\sum p_i h_i$	$\sum m_i h_i$	$\sum v_i h_i$	$\sum h_i$	$\sum h_i$	C
$\sum s_i k_i$	$\sum p_i k_i$	$\sum h_i k_i$	$\sum k_i^2$	$\sum t_i k_i$	$\sum m_i k_i$	$\sum v_i k_i$	$\sum k_i$	$\sum k_i$	D
$\sum s_i t_i$	$\sum p_i t_i$	$\sum h_i t_i$	$\sum k_i t_i$	$\sum t_i^2$	$\sum m_i t_i$	$\sum v_i t_i$	$\sum t_i$	$\sum t_i$	M
$\sum s_i m_i$	$\sum p_i m_i$	$\sum h_i m_i$	$\sum k_i m_i$	$\sum t_i m_i$	$\sum m_i^2$	$\sum v_i m_i$	$\sum m_i$	$\sum m_i$	F
$\sum s_i v_i$	$\sum p_i v_i$	$\sum h_i v_i$	$\sum k_i v_i$	$\sum t_i v_i$	$\sum m_i v_i$	$\sum v_i^2$	$\sum v_i$	$\sum v_i$	G
$\sum s_i$	$\sum p_i$	$\sum h_i$	$\sum k_i$	$\sum t_i$	$\sum m_i$	$\sum v_i$	n*	$\sum v_i$	N

Table 1.
Parameter matrices used in the least squares method.

increase after 2100. HadGEM2-ES is a second-generation global model developed by the Hadley Center, a research organization affiliated with the UK Met Office [12].

2.2 Method

2.2.1 Regression analysis

In this study, province-based regression equations were established using the least squares method (LSM) with sunflower yield values from the 29 provinces from 1985 through 2014 and the 7 selected climate parameters. After that, the potential impact of climate changes that are projected for the future periods (2016–2040, 2041–2070, and 2071–2099) on yield of sunflower has been put forward with using the generated high-rate regression equations and climate projection data (**Table 1**).

In the study, the regression analysis equation created by LSM was as follows:

$$y = As + Bp + Ch + Dk + Et + Fm + Gv + H \quad (1)$$

where the dependent variable $y = yield$. $A, B, C, D, E, F, G,$ and H are coefficients, and the independent variables were as follows:

s = monthly total sunshine duration (h)
p = monthly total precipitation (mm)
h = monthly average relative humidity (%)
k = number of days with daily average relative humidity >70%
t = monthly average temperature (°C)
m = number of days with daily maximum temperature >35°C
v = number of days with daily minimum temperature ≤−5°C

3. Results and discussion

This study was conducted in order to determine the possible effects of climate change on sunflower yield in 29 provinces where intensive sunflower cultivation has been conducted in Turkey. The periods of 2016–2040, 2041–2070, and 2071–2099 were determined as future periods.

3.1 Province-based yield-meteorological parameter regression analysis

In the first part of the study, it is aimed to determine the quality of the relationship between the variables using regression analysis. According to the results of multiple regression analyses using LSM between climate factors and yield, the rates of predicting province-based yield were very high.

The results indicate that predictability was between 0.38 and 0.50 in three provinces (Karaman, Kırşehir, and Uşak), it was between 0.51 and 0.69 in 11 provinces (Afyon, Aydın, Çorum, Diyarbakır, İstanbul, Kayseri, Kocaeli, Kütahya, Sakarya, Tekirdağ, and Tokat), and very high predictability (0.70–0.91) was determined in 15 provinces (Adana, Aksaray, Amasya, Ankara, Balıkesir, Bilecik, Bursa, Çanakkale, Edirne, Eskişehir, İzmir, Kırklareli, Konya, Osmaniye, and Samsun) (**Figure 2**).

3.2 Province-based sunflower yield change projections

In the second part of the study, multiple regression equations were used with the 20-km-resolution climate projection data from the HadGEM2-ES global climate model and RCP8.5 scenario. Yield estimation analyses were conducted for 2016–2040, 2041–2070, and 2071–2099. The obtained results were compared to the average yield values of 1985–2014, and the changes that may occur in sunflower yield were periodically examined.

3.2.1 Years 2016–2040

The results of the analyses conducted for 2016–2040 using the climate projections data in the regression equations predicted that sunflower yield would increase in 16 of the 29 provinces, decrease in 12 of the 29 provinces, and no change in 1 of the 29 provinces (**Figure 3**).

3.2.2 Years 2041–2070

Obtained results of the analysis conducted for 2041–2070 using the climate projections data in the regression equations, it is predicted that there will be an increase in sunflower yield in 17 of the 29 provinces and a decrease in yield in 12 of the 29 provinces (**Figure 4**).

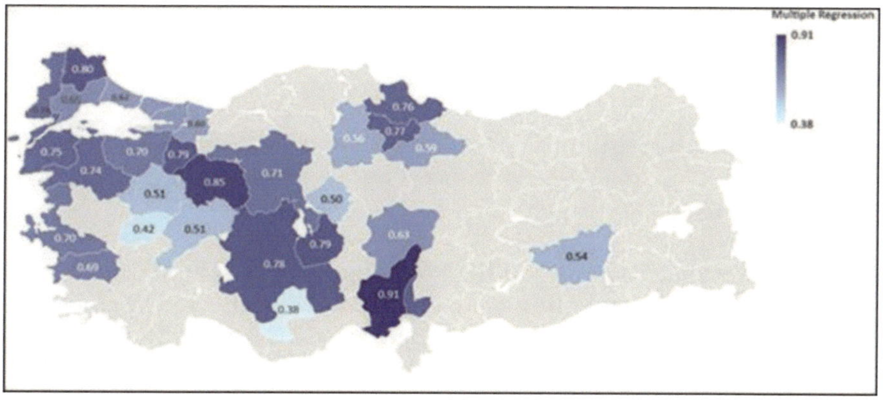

Figure 2.
Results of province-based yield meteorological parameter regression analysis.

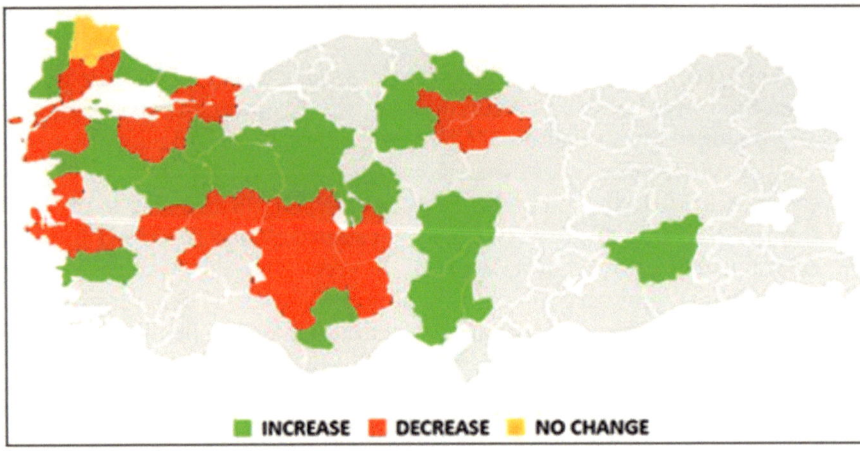

Figure 3.
Projected changes in sunflower yield for 2016–2040.

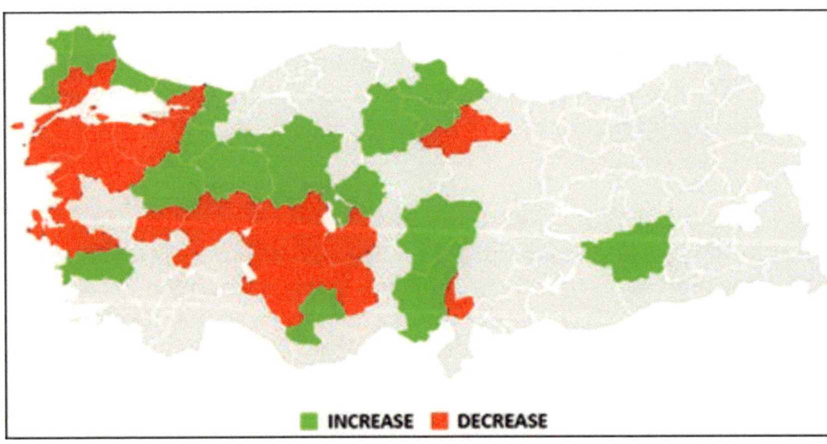

Figure 4.
Projected changes in sunflower yield for 2041–2070.

3.2.3 Years 2071–2099

The results of the analysis conducted for 2071–2099 using the climate projection data in the regression equations predicted that sunflower yield would increase in 16 of the 29 provinces, decrease in 12 of the 29 provinces, and no change in 1 of the 29 provinces (**Figure 5**).

3.3 Research results by region

3.3.1 Marmara region

Based on the evaluation results of the 10 provinces within the Marmara region, it is predicted that, in general, the region will be adversely affected by climate changes in the upcoming years, which would decrease average sunflower yields. The increases in the number of days with maximum temperatures >35°C would adversely affect the plant pollination period during the sunflower vegetation period throughout the region, which would have a negative impact on yield.

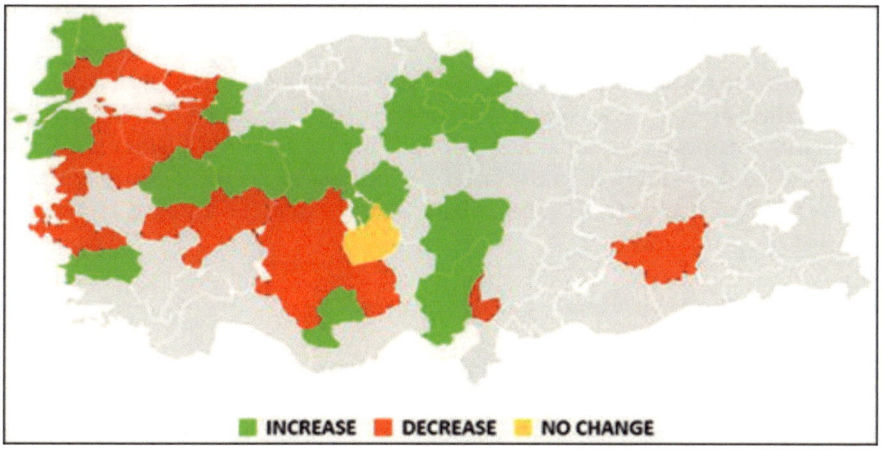

Figure 5.
Projected changes in sunflower yield for 2071–2099.

3.3.2 Central Anatolia region

The evaluation results within the seven provinces of the Central Anatolia region, it is predicted that the yield in Konya and Aksaray, where there is intensive sunflower cultivation, would decrease; on the other hand, it is predicted there will be increasing trend in other provinces of the region.

3.3.3 Aegean region

The evaluation results for the five provinces within the Aegean region suggest that the region would be adversely affected by climate changes, which would decrease the average sunflower yields within the region. It is expected that the climate changes during the studied future periods would positively affect sunflower yield in two of the provinces (Aydın and Kütahya) and negatively affect the yield in three of the provinces (Afyon, İzmir, and Uşak).

3.3.4 Black Sea region

The evaluation results of the four provinces within the Black Sea region suggest that the region would be positively affected by climate changes, especially during 2041–2070 and 2071–2099; therefore, there would be an increase in average sunflower yield. During these periods, the expected increase in the average temperature and changes in the average relative humidity are predicted to contribute positively to the yield.

3.3.5 Mediterranean region

Because of low number of provinces in the Mediterranean region in which sunflower was cultivated over the 30 years studied, the region could not be fully evaluated. Despite this, it is estimated that there will be increases in both provinces in the 2016–2040 period. In the second and third periods, excessive increases in both average temperatures and the number of days with daily maximum temperature >35°C are expected to adversely affect the sunflower yield.

3.3.6 Southeastern Anatolia region

Sunflower cultivation is not widespread in the Southeastern Anatolia region, and yield values are very low compared to those of the other regions. Nevertheless, the estimation results of Diyarbakır suggested that there would be increases in yield in 2016–2040 and 2041–2070. In the latter period, excessive increases in temperatures are expected to negatively affect yield, as is the case for the Mediterranean region.

3.4 Results based on the climate parameters

The following results are identified based on the climate parameters:

- The increase in average temperature generally contributes positively to yield.

- The average number of days with a minimum temperature ≤−5°C will not have a negative effect; however, any increase in the number of days would have a negative effect on yield.

- Limited increases in the number of days with a maximum temperature >35°C would have a positive effect on yield; however, excessive increases in those numbers would have a negative effect on yield, especially during 2041–2070 and 2071–2099.

- Decreases in average humidity and the number of days with an average humidity >70% are expected to have a positive effect on yield.

Because the average predictions for sunshine duration and total precipitation do not show significant province-based changes during all periods compared to that in 1985–2014, it can be suggested that they will not have negative effects on sunflower yield in the future periods; however, the rainfall parameter that was discussed in the study represents total for the growing season. Meteorological disasters, such as heavy rainfall and flooding, were excluded from the evaluations. As stated in the future climate projections, rather than increases in total rainfall, there are expected irregularities in the distribution of precipitation and meteorological disasters; therefore, it should be taken into consideration that precipitation parameter can have negative effect on yield in this way.

The results of the study showed that climate factors are not the sole determinant of sunflower yield but do have significant effects on production. According to the results of the analysis, it is concluded that especially temperature and humidity parameters have a significant effect on sunflower yield.

The yield estimations conducted with using the HadGEM2-ES global model and 20-km-resolution climate projections based on the RCP8.5 scenario indicate, regions in which sunflowers are cultivated will be affected by future climate changes. When regional comparisons were made in terms of sunflower cultivation based on climate changes, the Marmara and Aegean regions were found to be more sensitive to those changes. In terms of sunflower cultivation, decreases in yields in Konya and Aksaray in Central Anatolia were observed, whereas increases were predicted in other provinces within the same region. Similarly, in the study which examined the possible effects of climate change on oilseed cultivation in TR71 region, it was stated that temperature increases would increase water demand and disease probability and decrease in yield [7].

It is expected that the Mediterranean and Southeast Anatolia regions would be positively affected by climate changes in 2016–2040. It was shown that these two regions would be adversely affected by climate changes for 2041–2070 and 2071–2099. The Black Sea region is expected to be positively affected by climate changes, especially for 2041–2070 and 2071–2099.

4. Conclusion

In this study, a relationship between climate and yield was evaluated in areas in Turkey with intensive sunflower (oil type) production, and projections for changes in yield related to climate change were statistically analyzed. According to yield estimation projections, decreases are expected especially in the second half of the century. In the provinces of Tekirdağ and Konya, where intensive sunflower cultivation is conducted, the expectations of decreases in all future periods are remarkable. The results of this study can be compared with the yield projections using dynamic methods (crop simulation models), and the differences between the two methods can be determined. In addition, these results can be useful for determining the regions that should be encouraged in future product planning at the regional or national level, which can be conducted by taking into consideration the predicted changes in climate and sunflower yields.

Conflict of interest

The authors declare no conflict of interest.

Author details

Hudaverdi Gurkan[1], Yasin Ozgen[2], Nilgun Bayraktar[2], Huseyin Bulut[1] and Mustafa Yildiz[2]*

1 Turkish State Meteorological Service, Ankara, Turkey

2 Department of Field Crops, Faculty of Agriculture, Ankara University, Ankara, Turkey

*Address all correspondence to: myildiz@ankara.edu.tr

References

[1] TUIK Merkezi Dağıtım Sistemi [Internet]. 2019. Available from: https://biruni.tuik.gov.tr/medas/?kn=92&locale=tr [Accessed: 15 May 2019]

[2] TUIK [Internet]. 2019. Available from: http://www.tuik.gov.tr/PreTablo.do?alt_id=1001 [Accessed: 15 May 2019]

[3] Akçakaya A, Sümer UM, Demircan M, Demir Ö, Atay H, Eskioğlu O, et al. Turkey Climate Projections with New Scenario's and Climate Change—TR2015-CC. Ankara, Turkey: Turkish State Meteorological Service (TSMS); 2015. 149 p

[4] Demircan M, Gürkan H, Eskioglu O, Arabaci H, Coskun M. Climate change projections for Turkey: Three models and two scenarios. Turkish Journal of Water Science and Management. 2017;**1**(1):22-43

[5] Dellal İ, McCarl BA, Butt T. The economic assessment of climate change on Turkish agriculture. Journal of Environmental Protection and Ecology. 2011;**12**(1):376-385

[6] Soylu S, Sade B. A Research Project on the Impact of Climate Change on Agricultural Products. Konya, Turkey: Karapınar Ziraat Odası; 2012. 55 p

[7] Demir İ. Oilseed crop cultivation in TR71 region and effects of climate change. Turkish Journal of Agriculture - Food Science and Technology. 2013;**1**(2): 73-78

[8] Gurkan H. The effect of climatic factors and climate change on the yield of sunflower (*Helianthus annuus* L.) in Turkey [thesis]. Department of Field Crops, Graduate School of Natural and Applied Sciences, Ankara University; 2015

[9] Bulut H, Gürkan H, Arslan N. The effect of climatic factors and climate change on the yield of opium poppy (*Papaver somniferum* L.) in Turkey. In: 13. National Cultural Technical Congress; 12-14 April 2016; Antalya, Turkey; 2016. pp. 105-112

[10] Gurkan H, Bayraktar N, Bulut H, Kocak N, Eskioglu O, Demircan M. Analyzing of the potential impact of climate change on yield of sunflower (*Helianthus annuus* L.): Example of Marmara region. In: 12. National Agricultural Economy Congress; 25-27 May 2016; Isparta, Turkey; 2016. pp. 59-68

[11] Demircan M, Demir Ö, Atay H, Eskioğlu O, Yazıcı B, Gürkan H, Tuvan A, Akçakaya A. Climate change projections in Turkey with new scenarios. In: TUCAUM VIII. Geography Symposium; Turkey Geography Research Center, Ankara University; 23-24 October 2014; Ankara, Turkey; 2014. pp. 11-20

[12] TSMS. Turkey Climate Projections with New Scenario's, TR2013-CC. Ankara, Turkey: Turkish State Meteorological Service; 2013. 67 p

Chapter 4

Breeding for Biofortification Traits in Rice: Means to Eradicate Hidden Hunger

Vinay Sharma, Dinesh Kumar Saini, Ashish Kumar, Hari Kesh and Prashant Kaushik

Abstract

Rice (*Oryza sativa* L.) supplies nourishment to about half of the population of the world's inhabitants. Of them, more than 2 billion people suffer from 'hidden hunger' in which they are unable to meet the recommended nutrients or micronutrients from their daily dietary intake. Biofortification refers to developing micronutrient-rich diet foods using traditional breeding methods and modern biotechnology, a promising approach to nutrition enrichment as part of an integrated strategy for food systems. To improve the profile of rice grain for the biofortification-related traits, understanding the genetics of important biofortification traits is required. Moreover, these attributes are quantitative in nature and are influenced by several genes and environmental variables. In the course of past decades, several endeavours such as finding the important quantitative trait loci (QTLs) for improving the nutrient profile of rice seeds were successfully undertaken. In this review, we have presented the information regarding the QTLs identified for the biofortification traits in the rice.

Keywords: QTLs, biofortification, malnutrition, hidden hunger, marker-assisted breeding

1. Introduction

Rice (*Oryza sativa* L.) provides energy and nutrition to almost half of the world's population [1]. In most developing countries, especially in Asia, rice is consumed in significant quantities and is the main component diet. In the present scenario, high-yielding rice varieties are low in mineral elements. Milled or polished rice is not a significant source of any major mineral elements, and therefore, it cannot meet up with the recommended daily dietary intake for mineral elements. Moreover, around 792.5 million people across the world are malnourished, out of which 780 million people live in developing countries [2]. Thus, most rice-eating, resource-poor people in Southeast Asia, Africa, and Latin America suffer from chronic micronutrient malnutrition, often referred to as hidden hunger [3]. Protein-energy malnutrition affects 25% of children those with the dietary intake of predominantly rice, and staple crops have low levels of an essential amino acid [4]. Further, rice has relatively low (8.5%) protein content as compared to other cereals such as wheat, barley, and millets. Moreover, the average protein content in milled rice is around 7%. However, the total seed protein content of rice consists of 60–80% glutelin and

20–30% prolamin [5]. Interestingly, rice supplies about 40% of the total protein requirement of humans in developing countries [6].

Phytate is a crucial mineral storage compound in seed, with a mixed cation salt of phytic acid accounting for approximately 75% of total seed phosphorus content [7]. The significant portion of the phosphorus taken from the soil by plants is ultimately translocated to the seed and further synthesised into phytic acid. Phytate is vital for the development of seeds and also as an antioxidant, anticancer agent, lowering chronic disease rates, and preventing coronary disease [8]. Phytic acid is known as an anti-nutritional factor because it forms complexes in seeds with proteins and essential minerals such as Fe, Zn, and Ca [9] and leads to the impairment of the bioavailability of the same.

Mineral elements are critical for several cellular and metabolic activities [10]. Biofortification of staple crops provides a sustainable methodology to triumph over the mineral deficiency. Attempts were made for the development, release, and distribution of biofortified crops with the help of agronomic practices and biotechnological techniques and also by using plant breeding methods. Various old rice varieties with high grain iron and zinc content were screened, and breeding methods with improved agronomic characteristics combined the higher mineral characteristics. In 2013, the Bangladesh Rice Research Institute released zinc-enriched rice varieties (BRRIdhan 62, BRRIdhan 72, and BRRIdhan 64), claiming to contain 20–22 ppm of zinc in brown rice. An improved line (IR68144-3B-2-2-3) has been identified in India and Philippines in a cross between a high-yielding variety (IR72) and a large, traditional variety (ZawaBonday) with a top grain iron concentration about 21 ppm in brown rice [11].

Similarly, Jalmagna, a traditional variety with almost double the iron and zinc concentration of common rice variety, has been identified for further breeding programs to improve iron and zinc concentration by nearly 40 percent more than that of conventional rice variety [11]. ICAR-Indian Institute of Rice Research, Hyderabad, Telangana, developed biofortified pure line variety, DRR Dhan 45. It possesses high zinc (22.6 ppm) in polished grain. It has been released and notified in 2016 for Karnataka, Tamil Nadu, Andhra Pradesh, and Telangana. Its average grain yield is 50.0 q/ha. It matures in 125–130 days [12, 13]. Another pure line variety DRR Dhan 49 with high zinc (25.2 ppm) in polished grain is released and notified in 2018 for Gujarat, Maharashtra, and Kerala. Its average grain yield is 50.0 q/ha and matures in 125–130 days [13].

Mineral element accumulation in the grain is a complex process and is highly influenced by environmental factors. This resulted in less effective early-generation phenotypic selections for mineral grain elements and slowed progress in the breeding of biofortified rice varieties [14]. In-depth understanding of the genetic basis of mineral elements at the molecular level and the identification of significant effects of QTLs can help to speed up the development of biofortified rice varieties through marker-assisted breeding [15]. Rice is a model for cereal crops. Vast genomic resources are available, including genome-wide single nucleotide polymorphic (SNP) molecular markers and various advanced genomic platforms, to enable complex traits to be dissected at the molecular level [16]. Several studies to chart QTLs for biofortified traits include the use of introgression lines (ILs) [17] and double haploids (DHs) to uncover QTLs [18]. However, the stability of released genotypes is an important consideration to hope for a meticulous performance of released genotypes for stable produce for the farmers [19, 20]. Hence, molecular breeding approach for biofortification of crop offers a sustainable and long-term solution. Also, biofortified crops with increased bioavailability of essential protein, vitamins, and micronutrients are deployed to consumers through traditional farming and food trading practices, thus providing a feasible way to reach undernourished and low-income families with limited access to various diets, supplements, and fortified foods [21]. The common processes involved in the development of the biofortified rice variety (**Figure 1**).

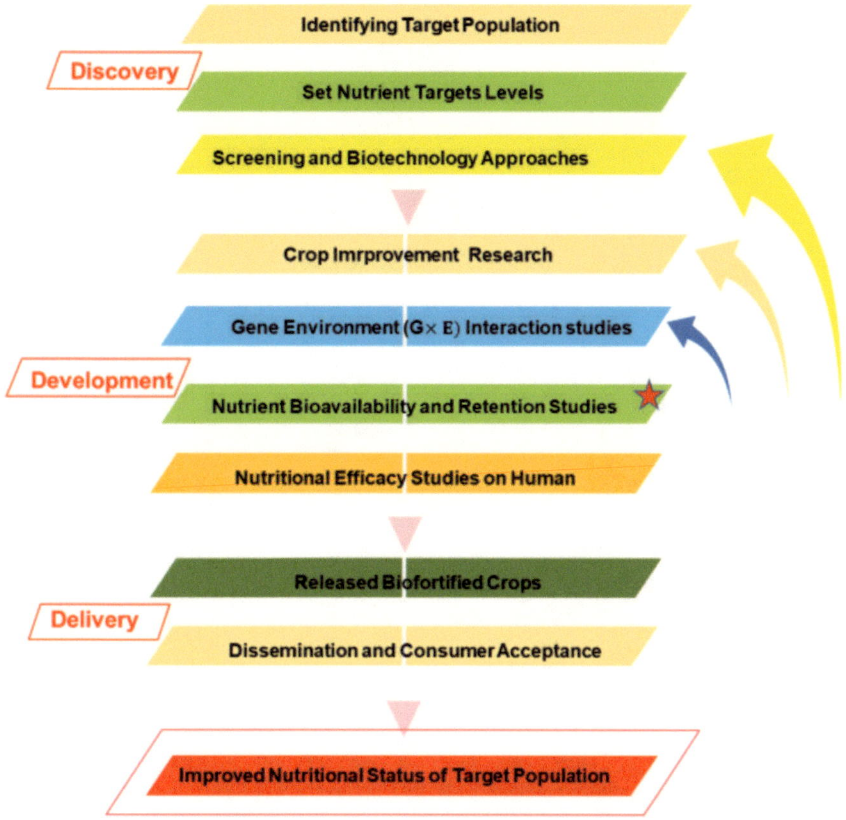

Figure 1.
Summary of the process involved in the biofortification of the rice.

2. Protein content in rice

Grain protein content (GPC) in rice is one of the major factors which decides the nutritional value of rice food and influences the palatability of cooked rice [22]. Rice's seed protein content consists of 60–80% glutelin and 20–30% prolamin, regulated by 15 and 34 genes, respectively [5]. It supplies about 40% of the protein to humans through diet in developing nations, and rice GPC quality is high, owing to lysine richness (3.8%) [6]. Improving GPC in rice grain is, therefore, a significant goal for plant breeders and biotechnologists. More than 20 QTL mapping studies have been conducted in the last two decades to explore the genetic base of the protein content in rice. Moreover, to our knowledge, more than 80 stable and consistent QTLs for GPC have been identified and mapped on all 12 chromosomes of rice, although most of them were mapped on chromosomes 1, 2, 6, 7, 10, and 11 (**Table 1**). For the first time, Tan et al. [28] mapped two QTLs, one in the interval of markers C952-Wx on chromosome 6, with the phenotypic variance explain (PVE) 13.0%, and the other one in the interval markers R1245-RM234 on chromosome 7 with PVE 6.0%. In another study, Aluko et al. [29] identified and mapped four QTLs among 312 DH lines derived from the BC_3F_1 of an interspecific cross of *O. sativa* × *O. glaberrima* explaining the phenotypic variance of 4.8–15.0%. Among the four QTLs, one QTL, pro6, was closely associated with Wx gene influencing rice quality. Thereafter, several studies have been conducted to map the QTLs regulating GPC in rice [26, 40–43].

Zheng et al. [39] employed unconditional and conditional QTL mapping methods to analyse the developmental behaviour of protein content and protein

Cross	Population type and size	No. of total QTLs	PVE range (additive effect QTLs)	Chromosomes/ chromosome arms	Marker intervals/nearest markers for major QTL (PVE)	References
Amino acid content						
Indica rice (Zhenshan 97) × Indica rice (Nanyangzhan)	RILs (190)	2 QTL clusters	4.05–33.3	1, 7	RM472-RM104 (Asp/Thr/Gly/Ala/Tyr/Pro/Lys/Ser/Glu/ Asp/Val/Met/Ile/Leu/Phe/His/Arg/Cys) (5.7–33.3)	[23]
Indica rice (Zhenshan 97) × Indica rice (Minghui 63)	RILs (241)	10 (His) + 8 (Arg)	12–35 (His); 16–33 (Arg)	1, 2, 3, 6, 7, 10, 11, 12 (His); 2, 3, 5, 6, 7, 10, 11, 12	R321–RM55 (12), RZ398–RM204 (12), RG101–G393 (15), C1003B–RG103 (15), RG118–C794 (20), RM53–RZ599 (22), RM258–RG561 (22), RG424–R2549 (23), RG528–RG128 (24), RM20b–C732 (35) [His]; C734b–RZ649 (16), R321–RM55 (18), RG424–R2549 (21),RM258–RG561 (21), R3203–RM20A (22), RM53–RZ599 (23), RG528–RG128 (23), RM20b–C732 (33)	[24]
Indica rice (Zhenshan 97) × Indica rice (Minghui 63)	RILs (241)	12	3.4–48.8	1, 11	R2632–C39 (Ser) (13.5), RG173–RM81A (Val) (14.5), RZ536–TEL3 (Met) (48.8)	[25]
Indica rice (Zhenshan 97B) × Indica rice (Delong 208)	RILs (188)	3 QTL clusters	4.2–31.7	1,7,9	RM328–RM107 (Asp/Thr/Ser/Gly/Val/Ile/Phe/Lys/Taa) (13.2), MRG186–MRG4499(Asp/Thr/Ser/Glu/Gly/ Ala/ Cys/Val/Met/Ile/Phe/Arg/Pro/Taa) (14.4–27.5), RM493–RM562 (Asp/Thr/Glu/Gly/Ala/Val/Leu/Phe/ Arg/Pro/Taa) (24.2 31.7)	[26]
O. sativa (Dasanbyeo) × *O. sativa* (TR22183)	RILs (172)	6	10.2–12.4	3	id3015453-id3016090 (Ala-10.2, Phe-10.6, Iso-11.2, Val-12.4, Leu-12.4), id3001422 fd10 (Lys-10.8)	[27]
Protein content						
Indica rice (Zhenshan 97) × Indica rice (Minghui 63)	RILs (238)	2	6.0–13.0	6, 7	C952-Wx (13)	[28]
Indica rice (Caiapo) × *Oryza glaberrima* (IRGC 103544)	DH lines (312)	4	4.8–15.0	1, 2, 6, 11	RM226–RM297 (15)	[29]
Indica rice (Gui630) × Japonica rice (02428)	DH lines (81)	5	6.9–35.0	1, 4, 5, 6, 7	C22-RG449d (16.5), ZG34B-G20 (22.5), RG435-RG172a (35.0)	[30]

Cross	Population type and size	No. of total QTLs	PVE range (additive effect QTLs)	Chromosomes/ chromosome arms	Marker intervals/nearest markers for major QTL (PVE)	References
O. sativa (V20A) × *O. glaberrima* (accession 103,544)	BC3(TC) F1 families (308)	1	9.0–10.0	8		[31]
Japonica rice (Moritawase) × Japonica rice (Koshihikari)	RILs (92)	3	2.3–16.3	2, 6, 9		[32]
Koshihikari/Indica rice (Kasalath)// Japonica rice (Koshihikari)	BILs (92)	2	14.3–14.8	6, 10	R1952 (14.3), R2447 (14.8)	[33]
Indica rice (Chuan) × Japonica rice (Nanyangzhan)	RILs (286)	2	2.69–4.50	6, 7		[34]
Indica rice (Xieqingzao B) × Indica rice (Milyang 46)	RILs (209)	5	3.9–19.3	3, 4, 5, 6, 10	RM251–RM282 (10.5), RM190–RZ516 (19.3)	[35]
Indica rice (Zhenshan 97) × Indica rice (Minghui 63)	RILs (241)	9	1.60–9.26	2, 3, 5, 6, 7, 10, 11, 12		[36]
Tongil variety (Samgang) × Japonica variety (Nagdong)	DH lines (120)	3	6.92–22.98	1, 11	RM287-RM26755 (21.21), 11,025-RM287 (22.98)	[37]
Japonica rice (Asominori) × Indica rice (IR24)	CSSLs (66)	9	3.0–53.7	1, 2, 3, 6, 8, 11	R1982 (10.4–14.2), XNpb113 (12.0–13.8), C1350 (23.6), G1149 (13.0–53.7)	[38]
Japonica rice (Asominori) × Indica rice (IR24)	RILs (71)	10	8.53–23.70	1, 3, 4, 6, 7, 8, 9, 10, 12	R265B-XNpb36 (10.50), C1003-C688 (12.67), XNpb212-G1318 (13.86), C606-XNpb238 (14.63), R1854-R2373 (15.65), XNpb24-C562 (17.60), XNpb338-C796 (19.59), R758-XNpb15 (19.74), XNpb268-R411 (23.70)	[39]
Indica rice (Zhenshan 97B) × Indica rice (Delong 208)	RILs (188)	2	7.2–25.9	1, 7	RM445–RM418 (25.9)	[26]

Cross	Population type and size	No. of total QTLs	PVE range (additive effect QTLs)	Chromosomes/ chromosome arms	Marker intervals/nearest markers for major QTL (PVE)	References
Koshihikari/Indica rice (Kasalath)// Japonica rice (Koshihikari)	BILs (182)	4	6.26–12.11	2, 3, 7, 10	R250-C746 (10.04), C16-C809 (11.07), C847-C596 (12.11)	[40]
Indica rice (Cheongcheong) × Indica rice (Nagdong)	DH lines (133)	1	39–41	2	RM12532–RM555 (39–41)	[41]
Japonica cultivar (CJ06) × Indica rice cultivar (TN1)	DH lines (116)	1	12.3–15.8	10	RM216-RM467 (12.3–15.8)	[42]
Indica rice (Cheongcheong) × Indica rice (Nagdong)	DH lines (133)	3	39–40	8,9,10	RM506-RM1235 (39), RM24934-RM25128 (40), RM219-RM23914 (40)	[43]
O. sativa (M201) × *O. sativa* (JY293)	RILs (234)	5 $	6.74–13.50	1, 2, 3, 4	RM423-RM6375 (11.72), GS3-SLAF13430 (13.50)	[44]
Japonica variety (Sasanishiki) × Indica variety (Habataki)	CSSLs (39)	1#	10.38–15.43	1	RM7124 (10.38–15.43)	[45]
Indica rice (Cheongcheong) × Indica rice (Nagdong)	DH lines (120)	1	14	7	RM8261 (14)	[46]
Naveen/O. sativa (ARC 10075)//O. sativa (Naveen)	BC3F5 lines (200)	3	6.70–17.35	1, 2, 7	CSCWR_Os01g02590__61041 (13.85), CSCWR_ Os02g10740_65058 (6.70–17.35)	[47]
Iron and Zinc						
Indica variety (IR64) × Japonica variety (Azucena)	DH lines (129)	GZn-2; GFe-3	q	1, 12; 2, 8, 12	RM235–RM17 (12.8), RM34–RM237 (15) [GZn]; RM270–RM17 (13.8), RM53–RM300 (16.5), RM137–RM325A (18.3) [GFe]	[18]
Indica cultivar (Zhengshan 97) × Indica cultivar (Minghui 63)	RILs (241)	GZn-3; GFe-2	5.3–18.61; 11.11–25.81	5, 7, 11; 1, 9	R3166-RG360 (12.34), C794-RG118 (18.61) [GZn]; C472-R2638 (11.11), RG236-C112 (25.81) [GFe]	[48]

Cross	Population type and size	No. of total QTLs	PVE range (additive effect QTLs)	Chromosomes/ chromosome arms	Marker intervals/nearest markers for major QTL (PVE)	References
O. sativa ssp. Indica (Teqing) × *O. rufipogon* Griff.	ILs (85)	GZn-2; GFe- 1	5–11; 7	5, 8; 2	RM152 (11) [Zn]	[17]
Indica rice (Bala) × Japonica rice (Azucena)	RILs (79)	GZn-4; GFe- 4	11.2–14.8; 9.7–21.4	6, 7, 10; 1, 3, 4, 7	G1082 (11.2), G20 (11.4), AB0601 (14.7), C223 (14.8) [GZn]; R1440 (15.5), C949 (16.2), R1618 (21.4) [GFe]	[49]
Indica cultivar (ZYQ8) x Japonica cultivar (JX17)	DH lines (127)	GZn-2	10.83–12.38	4, 6	CT206-G177 (10.83), RZ516-G30 (12.38) [GZn]	[34]
Indica rice (Madhukar) × Indica rice (Swarna)	RILs (168)	GZn-6; GFe-7	29–35; 69–71	3, 7, 12; 1, 5, 7, 12	RM501–OsZip2 (29), RM7–RM517 (31), RM260–RM7102 (34), RM234–RM248 (35), RM248–RM8007 (35), RM17–RM260 (35) [GZn]; RM243–RM488 (69), RM488–RM490 (69.2), RM574–RM122 (69.2), RM234–RM248 (69), RM248–RM8007 (69), RM17–RM260 (71), RM 260–RM7102 (71) [GFe]	[50]
Indica rice (PAU201) x Indica rice (Palman 579)	F2 (247)	GZn-3; GFe- 8	4.7–19.1; 2.4–26.8	2, 10; 2, 3, 7, 10, 12	8RM474–RM184 (19.1) [Zn]; RM491–RM519 (16.9), RM228–RM496 (18.1), RM53–RM521 (21.4), RM221–RM208 (26.8)	[51]
Indica cultivar (Ce258) x Japonica breeding line (IR75862) and Indica cultivar (ZGX1) x Japonica breeding line (IR75862)	BILs (200 and 201)	GZn-4; GFe-1	2–24.4; 10.2–18.3	3, 6, 7, 8; 6, 11	RM293–RM85 (11.1–14.4), RM407–RM152 (11.2–18.0) [GZn]; RM3–RM340 (10.2–18.3) [GFe]	[44]
Indica cultivar (Swarna) X Japonica rice (Moroberekan)	RILs (60)	GFe-1	39	1	RM490-RM5 (39)	[52]
O. sativa (XB) × *O. rufipogon* (accession of DWR)	BILs (202)	GZn-6; GFe-3	5.3–11.8; 6.1–28.2	3, 4, 6, 7, 10, 12; 3, 6, 9	RG172-RM340 (11.8) [GZn]; RG123-RG172 (16.7), RG510-RZ251 (28.2)	[53]
O. sativa (Nipponbare)/*O. meridionalis* (W1627)//Nipponbare	BRILs (151)	GZn-4	15.0–21.9	2, 9, 10	RM171-RM590 (15.0), RM573 (15.2), RM6 (17.6), RM24085-RM566 (21.9)	[54]

Cross	Population type and size	No. of total QTLs	PVE range (additive effect QTLs)	Chromosomes/ chromosome arms	Marker intervals/nearest markers for major QTL (PVE)	References
Indica cultivar (PSBRc82) x Korean rice (Joryeongbyeo) and PSBRc82 x Indica breeding line (IR69428)	DH lines (130 and 97)	GZn-8; GFe-1	7.5–22.8; 9.4	2, 3, 6, 8, 11, 12; 4	2,140,834–2,147,095 (10.3), 13,048,465–13,057,679 (12.3), 8,803,052–8,832,534 (14.3), 6,025,827–6,047,367 (15.3), 606,341- id6006214 (16.1), 2,110,566-id2009463 (17.3), 2,783,884–2,785,595 (20.3), 10,858,811-id11000778 (22.8) [GZn]	[55]
Indica cultivar (IR64) × Breeding line (IR69428) and Indica cultivar (BR29) × Breeding line (IR75862)	DH lines (111 and 146)	GZn-8	8.6–27.7	2, 3, 5, 7, 8, 9, 11	wd9002310–9,831,169 (10.3), 5,645,339–5,648,872 (11.5), 2,048,774–2,054,640 (12.2), 3,538,410–3,548,096 (12.2), 7,062,019–7,089,136 (12.6), 5,027,770–5,077,125 (18.4), 10,907,196-id11001107 (27.7) [GZn]	[56]
Indica rice (PAU201) x Indica rice (Palman)	F4 Population (579)	GZn-1; GFe-5	25; 34.6–95.2	6; 5, 7, 9	RM585-RM3 (25) [GZn]; RM2488-RM440 (64.1), RM440-RM31 (95.2), RM440-RM31 (95.2), RM432-RM429 (95.2), RM566-RM434 (36.6) [GFe]	[57]

Table 1.
List of QTLs identified for biofortification traits in rice.

index in rice. At four stages of grain filling, viz. 7, 14, 21, and 28 DAF, they mapped 10 unconditional QTLs and 6 conditional QTLs, explaining 8.53–19.59% and 8.76–23.70% of PVE for GPC, respectively, and 11 unconditional QTLs and 9 conditional QTLs explaining 7.46–16.97% and 7.46–18.88% of PVE for protein index, respectively. A strategy to detect more QTLs for rice grain quality within subpopulations [44]. Xu et al. [58] detected a total of 29 QTLs in the whole population and 10 QTLs in the two subpopulations for 7 traits, 4 of which (1 qPRO3.1 for protein content) were detected in the entire population but the remaining 6 QTLs were not. These six QTLs with minor effects might have been covered by the Wx locus when mapped in the whole population. In addition to usual biparental populations such as recombinant inbred lines, backcross inbred lines, and doubled haploid lines, advanced population, i.e. chromosome segment substitution line (CSSL) populations, has also been employed [45]. Yang et al. [45] used a CSSL population derived from the cross of a Japonica variety (Sasanishiki) with Indica variety (Habataki) and identified a total of seven QTLs in three environments, although only one QTL (qPC-1) was detected across three environments explaining 10.38–15.43% of PVE. Furthermore, they developed F_2 and F_3 segregating populations from the cross between a CSSL with low PC, SL402, harbouring qPC-1 and Sasanishiki, and delimited the region of qPC-1 to a 41-kb on chromosome 1. These results may be helpful to introgress the QTL for GPC into rice cultivars using marker-assisted selection. In one study, Bruno et al. [46] observed compromised heritability percentage for protein while higher heritability percentage for the amylose content in a DH population derived from a cross between Cheongcheong and Nagdong. They mapped a QTL for GPC on chromosome 7 linked with the marker RM8261, explaining 14% of PVE.

As has been shown by previous studies, identification of robust QTLs for GPC in rice grains has been restricted because of lack of appropriate donors, non-utilisation of high-throughput phenotyping and genotyping platforms, and high genotype × environment (G × E) interaction. To overcome these restrictions, recently Chattopadhyay et al. [47] genotyped a BC_3F_4 mapping population derived from the cross between grain protein donor, ARC10075 and high-yielding cultivar Naveen, using 40 K Affymetrix custom SNP array, and identified three stable QTLs (viz. qGPC1.1, qSGPC2.1, and qSGPC7.1) for GPC explaining 13, 14, and 7.8% of PVE, respectively. QTLs identified in this study can be useful to improve the nutritional quality of rice grain. The closely linked markers that flanked the identified QTLs can be used to aid quality selection in breeding programs. And the results of the coincidence between the QTL detected, and the loci involved in protein biosynthesis pathways, might be helpful for gene cloning by the candidate gene method.

3. Amino acid content

In addition to GPC, improvement in the amino acid composition is important to meet the food demands of a growing global population. A major function of proteins in nutrition is to supply adequate amounts of required amino acids [59, 60]. Depending on requirement and availability in animal metabolic processes, essential amino acids cannot be synthesised by animals, but play a crucial role in metabolism [61]. Therefore, improving amino acid content in rice grain is an important objective. Several studies using the linkage mapping approach with various mapping populations have provided useful genetic information for improving the amino acid composition (AAC) in rice grains. Wang et al. [23] identified 18 chromosomal regions for 19 components of AAC in 2 years, viz. 2002 and 2004. They found a total of 10 QTL clusters in 2002 and 6 in 2004.

Interestingly, they also detected a wide coincidence between the QTLs and the loci involved in amino acid metabolism pathways, including N assimilation, transfer and protein biosynthesis. In a similar study, Zhong et al. [26] reported 48 and 64 QTLs, each contributing 4.0–43.7% to the total phenotypic variance, in 2004 and 2005, respectively. They also reported good coincidence between the detected QTL and the loci involved in amino acid metabolism pathways in nitrogen assimilation and transport, or protein biosynthesis. In another study, Zheng et al. [24] mapped a total of 10 QTLs explaining 12–35% of PVE for histidine on chromosomes 1, 2, 3, 6, 7, 10, 11, and 12 and 8 QTLs explaining 16–33% of PVE for arginine on chromosomes 2, 3, 5, 6, 7, 10, 11, and 12. All QTLs showed significant additive effects from the triploid endosperm and diploid maternal plant, while two QTLs for histidine and two for arginine content also showed significant dominant main effects from the triploid endosperm. Various interactions between QTLs and the environment were detected for five QTLs associated with histidine content and two QTLs associated with arginine contents. QTLs associated with amino acid contents and linked/flanking markers are summarised in **Table 1**. Recently, Yoo [27] mapped a total of six main-effect QTLs located on chromosome 3, contributing 10.2–12.4% PVE for the content of six amino acids. The QTL cluster (qAla3, qVal3, qPhe3, qIle3, and qLeu3) in the interval of markers id3015453 and id3016090 was found to be associated with the contents of five amino acids and accounted for PVE from 10.2 to 12.8%. Although they also detected 26 digenic interactions for the content of 7 amino acids, viz. Asp, Ser, His, Gly, Arg, Ala, and Tyr, involving 25 loci distributed on the 9 chromosomes, but they did not find any interaction for the other 9 amino acids. Therefore, these identified QTL results will be useful to find the candidate genes and favourable alleles for the enrichment of nutritional value in rice grain.

4. Zn and Fe contents in rice

Zn deficiency in grown-up children and adolescent males causes retarded growth and dwarfism, retarded sexual development, impaired sense of taste and poor appetite, and mental lethargy [62]. Several roles of zinc are found to be involved in an abundant number of proteins in biological systems to maintain their structural stability function. It has been found that Zn is essential for gene regulation and expression under stress conditions and is therefore required for protection against infections and disease [63]. Likewise, iron has so many vital functions in the body like as a carrier of oxygen to the tissues from the lungs [64].

In last two decades, more than 80 QTLs have been identified and mapped on all 12 chromosomes of rice for zinc and iron contents using various mapping populations derived from different intraspecific and interspecific crosses. QTLs associated with zinc and iron contents and linked/flanking markers are summarised in **Table 1**. As per our knowledge, for the first time, Stangoulis et al. [18] mapped two QTLs for Zn and three QTLs for Fe on chromosomes 1, 2, 8, and 12 explaining 12.8–15% and 13.8–18.3% of PVE, respectively. Besides, Garcia- Oliveira et al. [17] detected one major effect QTL explaining the most significant proportion of PVE (11–19%) for zinc, flanking SSR marker RM152 on chromosome 8. In other various studies, several QTLs have been reported which explained a large amount of PVE either for zinc or for both iron and zinc contents [34, 48–52].

Ishikawa et al. [53] mapped four QTLs on chromosomes 2, 9, and 10 explaining 15.0–21.9% of PVE for grain zinc content using backcross recombinant inbred lines (BRILs) derived from *O. sativa* 'Nipponbare' and *O. meridionalis* W1627. Further, they fine-mapped QTL (named qGZn9) present on chromosome 9 and identified two tightly linked loci, qGZn9a (candidate region-190 kb) and qGZn9b (950 kb). They also showed the association of wild chromosomal segment covering qGZn9a with fertility reduction, and hence they recommended the use of qGZn9b as a valuable allele for breeding rice with high Zn in the grains. In another study, Swamy et al. [55] identified 20 QTLs for agronomic traits and total 59 QTLs for several biofortification traits including 8 QTLs for grain zinc and one QTL for grain iron, mapped on chromosomes 2, 3, 4, 6, 8, 11, and 12. They also detected eight epistatic interactions for Zn, Cu, Mg, and Na in a double haploid population.

Furthermore, they identified several candidate genes near grain zinc QTL (OsNRAMP, OsNAS, OsZIP, OsYSL, OsFER, and OsZIFL family), which may be useful for marker-assisted breeding for this important trait. Recently in 2019, two critical studies were conducted; in the first study, Descalsota-Empleo et al. [55] phenotyped two DH populations at two seasons and genotyped with a 6 K SNP chip and identified a total of 15 QTLs for agronomic traits and 50 QTLs for grain element concentration including 8 QTLs explaining 8.6–27.7% PVE for grain zinc. They also analysed the combined effect of QTL in both populations. Among the single-QTL lines, those with qZn9.1 showed highest mean grain Zn of 18.1 and 19.1 mg kg^{-1} in two consecutive seasons, respectively. They reported an increase in the content of zinc with the increase in number of QTLs and observed highest grain Zn of 28.2 and 24.3 mg kg^{-1} in two seasons, respectively, in four QTL lines (qZn2.1 + qZn5.1 + qZn5.1 + qZn11.1). Their results showed the possibility of QTL pyramiding for improving the zinc content in rice. In another study, Kumar et al. [57] detected one QTL for Zn and five QTLs for Fe having PVE 25% and 34.6–95.2%, respectively, using F_4 population (579 individuals) derived from a cross between PAU 201 and Palman. These identified QTLs can significantly enhance the efficacy of breeding programs to improve the Zn and Fe density in rice. The Zinc fortified rice varieties are released globally (**Figure 2**).

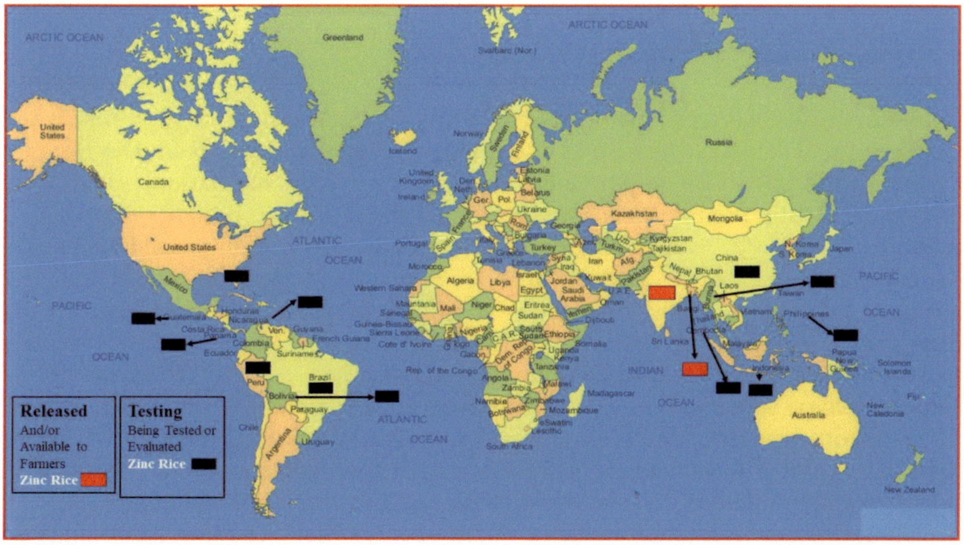

Figure 2.
Map showing countries where zinc-biofortified rice varieties are released and being tested (information taken from HarvestPlus).

5. Phytic acid

Phytic acid is an essential constituent in staple foods like legume and cereals, which has been of much concern [57]. In addition to its beneficial effect on human health, it has some anticancer and antioxidant functions and prevents coronary disease, and it is well known that phytic acid acts as strong chelating agent of mineral nutrients such as Ca, Zn, and Fe [65]. It has been seen that due to the presence of complex of phytic acid, in the form of phytate, there is a significant reduction found in bioavailability of nutrient elements [66]. It seems reasonable to control phytic acid contents in edible parts of crops to a level in which the medical and health functions of the food may be maintained and bioavailability of minerals is not much altered [67].

Liu et al. [68] assayed 72 cultivars for protein content and phytic acid and reported a wide range for phytic acid ranging from 0.685 to 1.03%, with an average of 0.873%. Interestingly, grain phytic acid and protein content were not correlated, which suggests the possibility of breeding rice for phytic acid and high protein content. Furthermore, they also reported a significant effect of varieties, locations, and their interactions on phytic acid content, with the location having the most considerable impact which suggests the necessity of multi-environment trials for the accurate evaluation of rice germplasm for phytic acid content.

Although sufficient genetic variation for phytic acid has been reported in various studies [68, 69], unfortunately, only one study has been conducted to map the QTLs for phytic acid in rice [18]. Stangoulis et al. [18] identified two QTLs explaining 15.4–24.3% of PVE for grain phytates from an IR64 × Azucena double haploid population. One common QTL for phytate and total P concentrations on chromosome 5 with the (high concentration) allele contributed from Azucena was identified. Furthermore, it was reported that Fe, Zn, and Mn contents in grains have different genetic regulation because the QTLs of phytate were not located on the same chromosomal regions as those found for Fe, Zn, and Mn [18]. So, there is a great possibility to find segregants having a low level of phytic acid and high level of Fe, Zn, and Mn content. Use of molecular marker in the breeding and selection to reduce grain phytic acid and improving the nutritional value of cereal grains.

6. Conclusions and future prospects

Biofortification is a promising, cost-effective, agricultural strategy to improve the nutritional status of the world's undernourished populations. Strategies for biofortification based on crop breeding, targeted genetic manipulation, and/or mineral fertiliser application have great potential to address human mineral malnutrition [70–72]. Developing biofortified food crops with improved nutrient content such as increased content of iron, zinc, Se, and provitamin A provides adequate levels of these and other such micronutrients that are often lacking in developed and developing diets. International initiatives, such as the CGIAR centres in collaboration with HarvestPlus and national initiatives, serve as pillars for achieving these objectives. These efforts have resulted in crops with the potential to increase both quantities and bioavailability of essential mineral elements in human diets, particularly in elementary cereal crops such as rice, wheat, maize, cassava, beans, and sweet potatoes. However, crop biofortification is a challenging task. Collaboration between plant breeders, nutritionists, genetic engineers, and molecular biologists is essential to achieving this. Breeding approaches are generalised and easy to accept and have been used to improve food nutritional qualities sustainably. Although greater emphasis is placed on molecular breeding-based approaches of which success rates are much higher as transgenically fortified crop plants, it faces

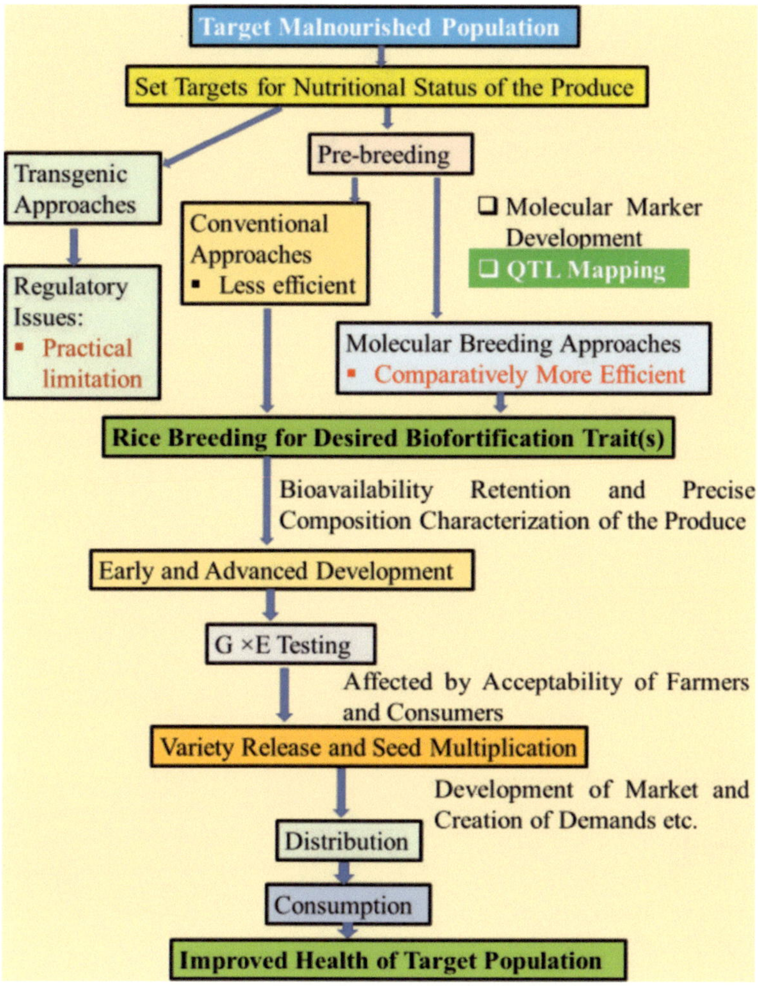

Figure 3.
Flowchart showing development and release of a biofortified variety and its acceptance by farmers and consumers.

hurdles due to consumer acceptance and costly and time-consuming regulatory approval processes adopted by different countries. Biofortified crops have a very bright future in addition to these challenges, as they have the potential to eliminate micronutrient malnutrition among billions of poor people, particularly in developing countries. Overall developmental process of the biofortified rice variety are presented in **Figure 3**.

Conflict of interest

The authors declare no conflict of interest.

Author details

Vinay Sharma[1], Dinesh Kumar Saini[2], Ashish Kumar[3], Hari Kesh[4] and Prashant Kaushik[5,6]*

1 International Crops Research Institute for the Semi-Arid Tropics, India

2 Department of Plant Breeding and Genetics, Punjab Agricultural University, India

3 Division of Genetics, ICAR-Indian Agricultural Research Institute, India

4 Department of Genetics and Plant Breeding, CCS Haryana Agricultural University, Hisar, India

5 Instituto de Conservación y Mejora de la Agrodiversidad Valenciana, Universitat Politècnica de València, Valencia, Spain

6 Nagano University, Ueda, Japan

*Address all correspondence to: prakau@doctor.upv.es

References

[1] GRiSP (Global Rice Science Partnership). Rice Almanac. 4th Ed. Los Baños (Philippines): International Rice Research Institute; 2013

[2] McGuire S, FAO, IFAD, WFP. The state of food insecurity in the world 2015: Meeting the 2015 international hunger targets: Taking stock of uneven progress. Rome: FAO. Advances in Nutrition. 2015;**6**:623

[3] Muthayya A, Rah JH, Sugimoto JD, Roos FF, Kraemer K, Black RE. The global hidden hunger indices and maps: An advocacy tool for action. PLoS One. 2013;**8**(6):e67860

[4] Gearing ME. Good as gold: Can golden rice and other biofortified crops prevent malnutrition. Science in the News Harvard University. 2015. Available form: http://sitn.hms.harvard.edu

[5] Xu JH, Messing J. Amplification of prolamin storage protein genes in different subfamilies of the Poaceae. Theoretical and Applied Genetics. 2009;**119**:1397

[6] Shobharani N, Pandey MK, Prasad GSV, Sudharshan I. Historical significance in grain quality features and precision breeding for improvement of export quality basmati varieties in India. Journal of Crop Science. 2006;**1**:29

[7] Raboy V. Approaches and challenges to engineering seed phytate and total phosphorus. Plant Science. 2009;**177**:281

[8] Gemede HM. Potential health benefits and adverse effects associated with phytate in foods. Food Science and Quality Management. 2014;**27**:45

[9] Tamanna S, Sayma P, Sanjay K, Alak KD, Aysha FM, Ali S, et al. Content of some minerals and their bioavailability in selected popular rice varieties from Bangladesh. International Journal of Current Microbiology and Applied Sciences. 2013;**2**:35

[10] Institute of Medicine Food and Nutrition Board (IMFNB). Dietary Reference Intakes for Vitamin a, Vitamin K, Arsenic, Boron, Chromium, Copper, Iodine, Iron, Manganese, Molybdenum, Nickel, Silicon, Vanadium and Zinc. Washington: National Academy Press; 2001

[11] Gregorio GB, Senadhira D, Htut H, Graham RD. Breeding for trace mineral density in rice. Food and Nutrition Bulletin. 2000;**21**:382

[12] Yadava DK, Choudhury PR, Hossain F, Kumar D. Biofortified Varieties: Sustainable Way to Alleviate Malnutrition. New Delhi: Indian Council of Agricultural Research; 2017

[13] Yadava DK, Hossain F, Choudhury PR, Kumar D, Singh AK. Biofortification of crops: A sustainable tool for nutritional security in India. Indian Farming. 2018;**68**:37-42

[14] Perez-de-Castro AM, Vilanova S, Canizares J, Pascual L, Blanca JM, Diez MJ, et al. Application of genomic tools in plant breeding. Current Genomics. 2012;**13**:179

[15] Zhang M, Pinson SR, Tarpley L, Huang XY, Lahner B, Yakubova E, et al. Mapping and validation of quantitative trait loci associated with concentrations of 16 elements in unmilled rice grain. Theoretical and Applied Genetics. 2014;**127**:137

[16] McCouch SR, Wright MH, Tung CW, Maron LG, McNally KL, Fitzgerald M, et al. Open access resources for genome-wide association mapping in rice. Nature Communications. 2016;**7**:10532

[17] Garcia-Oliveira AL, Tan L, Fu Y, Sun C. Genetic identification of quantitative trait loci for contents of mineral nutrients in rice grain. Journal of Integrated Plant Biology. 2009;**51**:84

[18] Stangoulis JC, Huynh BL, Welch RM, Choi EY, Graham RD. Quantitative trait loci for phytate in rice grain and their relationship with grain micronutrient content. Euphytica. 2007;**154**:289

[19] Jain BT, Sarial AK, Prashant K. Stability analysis utilizing AMMI model and regression analysis for grain yield of basmati rice (*Oryza sativa* L.) genotypes. JEBAS. 2018;**6**:522

[20] Jain B, Sarial A, Kaushik P. Understanding GxE interaction of elite basmati Rice (*Oryza sativa* L.) genotypes under north Indian conditions using stability models. Applied Ecology and Environmental Research. 2019;**17**:5863

[21] Hefferon KL. Can biofortified crops help attain food security? Current Molecular Biology Research. 2016;**2**:180

[22] Matsue Y, Odahara K, Hiramatsu M. Differences in amylose content amylographic characteristics and storage proteins of grains on primary and secondary rachis branches in rice. Japanese Journal of Crop Science. 1995;**64**:601

[23] Wang L, Zhong M, Li X, Yuan D, Xu Y, Liu H, et al. The QTL controlling amino acid content in grains of rice (*Oryza sativa*) are co-localized with the regions involved in the amino acid metabolism pathway. Molecular Breeding. 2008;**21**:127

[24] Zheng X, Jian-Guo WU, LOU X-Y, Hai-Ming XU, SHI C-H. QTL analysis of maternal and endosperm genomes for histidine and arginine in rice (*Oryza sativa* L.) across environments. Acta Agronomica Sinica. 2008;**34**:369

[25] Lu K, Li L, Zheng X, Zhang Z, Mou T, Hu Z. Genetic dissection of amino acid content in rice grain. Journal of Food Science & Agriculture. 2009;**89**:2377

[26] Zhong M, Wang LQ, Yuan DJ, Luo LJ, Xu CG, He YQ. Identification of QTL affecting protein and amino acid contents in rice. Rice Science. 2011;**18**:187

[27] Yoo SC. Quantitative trait loci controlling the amino acid content in rice (*Oryza sativa* L.). Journal of Plant Biotechnology. 2017;**44**:349

[28] Tan YF, Sun M, Xing YZ, Hua JP, Sun XL, Zhang QF, et al. Mapping quantitative trait loci for milling quality protein content and color characteristics of rice using a recombinant inbred line population derived from an elite rice hybrid. Theoretical and Applied Genetics. 2001;**103**:1037

[29] Aluko G, Martinez C, Tohme J, Castano C, Bergman C, Oard JH. QTL mapping of grain quality traits from the inter-specific cross between *Oryza sativa* × Oryza glaberrima. Theoretical and Applied Genetics. 2004;**109**:630

[30] Hu ZL, Li P, Zhou MQ, Zhang ZH, Wang LX, Zhu LH, et al. Mapping of quantitative trait loci (QTLs) for rice protein and fat content using doubled haploid lines. Euphytica. 2004;**135**:47

[31] Li J, Xiao J, Grandillo S, Jiang L, Wan Y, Deng Q, et al. QTL detection for rice grain quality traits using an inter-specific backcross population derived from cultivated Asian (*O. sativa* L.) and African (*O. glaberrima* S.) rice. Genome. 2004;**47**:697

[32] Wada T, Uchimura Y, Ogata T, Tsubone M, Matsue Y. Mapping of QTLs for physicochemical properties in japonica rice. Breeding Science. 2006;**56**:253

[33] Takeuchi Y, Nonoue Y, Ebitani T, Suzuki K, Aoki N, Sato H, et al. QTL detection for eating quality including glossiness stickiness taste and hardness of cooked rice. Breeding Science. 2007;**57**:231

[34] Zhang X, Zhang G, Guo L, Wang H, Zeng D, Dong G, et al. Identification of quantitative trait loci for Cd and Zn concentrations of brown rice grown in Cd-polluted soils. Euphytica. 2011;**180**:173

[35] Lou J, Chen L, Yue G, Lou Q, Mei H, Xiong L, et al. QTL mapping of grain quality traits in rice. Journal of Cereal Science. 2009;**50**:145

[36] Yu YH, Li G, Fan YY, Zhang KQ, Min J, Zhu ZW, et al. Genetic relationship between grain yield and the contents of protein and fat in a recombinant inbred population of rice. Journal of Cereal Science. 2009;**50**:121

[37] Shi CH, Shi Y, Lou XY, Xu HM, Zheng X, Wu JG. Identification of endosperm and maternal plant QTLs for protein and lysine contents of rice across different environments. Crop & Pasture Science. 2009;**60**:295

[38] Qin Y, Kim SM. Sohn JK. QTL analysis of protein content in double-haploid lines of rice. Korean Journal of Crop Science. 2009;**54**:165

[39] Zheng L, Zhang W, Chen X, Ma J, Chen W, Zhao Z, et al. Dynamic QTL analysis of rice protein content and protein index using recombinant inbred lines. Journal of Plant Biology. 2011;**54**:321

[40] Zheng L, Zhang W, Liu S, Chen L, Liu X, Chen X, et al. Genetic relationship between grain chalkiness protein content and paste viscosity properties in a backcross inbred population of rice. Journal of Cereal Science. 2012;**56**:153

[41] Lee GH, Yun BW, Kim KM. Analysis of QTLs associated with the rice quality related gene by double haploid populations. International Journal of Genomics. 2014

[42] Leng Y, Xue D, Yang Y, Hu S, Su Y, Huang L, et al. Mapping of QTLs for eating and cooking quality-related traits in rice (*Oryza sativa* L.). Euphytica. 2014;**197**:99

[43] Yun BW, Kim MG, Handoyo T, Kim KM. Analysis of rice grain quality-associated quantitative trait loci by using genetic mapping. Asian Journal of Plant Sciences. 2014;**5**(09):1125

[44] Xu F, Sun C, Huang Y, Chen Y, Tong C, Bao J. QTL mapping for rice grain quality: A strategy to detect more QTLs within sub-populations. Molecular Breeding. 2015;**35**:105

[45] Yang Y, Guo M, Li R, Shen L, Wang W, Liu M, et al. Identification of quantitative trait loci responsible for rice grain protein content using chromosome segment substitution lines and fine mapping of qPC-1 in rice (*Oryza sativa* L.). Molecular Breeding. 2015;**35**:130

[46] Bruno E, Choi YS, Chung IK, Kim KM. QTLs and analysis of the candidate gene for amylose protein and moisture content in rice (*Oryza sativa* L.). 3 Biotechnology. 2017;**7**:40

[47] Chattopadhyay K, Behera L, Bagchi TB, Sardar SS, Moharana N, Patra NR, et al. Detection of stable QTLs for grain protein content in rice (*Oryza sativa* L.) employing high throughput phenotyping and genotyping platforms. Science Reporter. 2019;**9**:3196

[48] Lu K, Li L, Zheng X, Zhang Z, Mou T, Hu Z. Quantitative trait loci controlling Cu, Ca, Zn, Mn and Fe content in rice grains. Journal of Genetics. 2008;**87**:305

[49] Norton GJ, Deacon CM, Xiong L, Huang S, Meharg AA, Price AH. Genetic mapping of the rice genome in leaves and grain: Identification of QTLs for 17 elements including arsenic, cadmium, iron and selenium. Plant and Soil. 2010;**329**:139

[50] Anuradha K, Agarwal S, Rao YV, Rao KV, Viraktamath BC, Sarla N. Mapping QTLs and candidate genes for iron and zinc concentrations in unpolished rice of Madhukar× Swarna RILs. Genetics. 2012;**508**:233

[51] Kumar J, Jain S, Jain RK. Linkage mapping for grain iron and zinc content in F_2 population derived from the cross between PAU 201 and Palman 579 in rice (*Oryza sativa* L.). Cereal Research Communications. 2014;**42**:389

[52] Indurkar AB, Majgahe SK, Sahu VK, Vishwakarma A, Premi V, Shrivastatva P. Identification characterization and mapping of QTLs related to grain Fe, Zn and protein contents in Rice (*Oryza sativa* L.). Electronic Journal of Plant Breeding. 2015;**6**:1059

[53] Hu BL, Huang DR, Xiao YQ, Fan YY, Chen DZ, Zhuang JY. Mapping QTLs for mineral element contents in brown and milled rice using an *Oryza sativa* × *Oryza rufipogon* backcross inbred line population. Cereal Research Communications. 2016;**44**:57

[54] Ishikawa R, Iwata M, Taniko K, Monden G, Miyazaki N, Orn C, et al. Detection of quantitative trait loci controlling grain zinc concentration using Australian wild rice (*Oryza meridionalis*) a potential genetic resource for biofortification of rice. PLoS One. 2017;**12**:e0187224

[55] Swamy BM, Descalsota GIL, Nha CT, Amparado A, Inabangan-Asilo MA, Manito C, et al. Identification of genomic regions associated with agronomic and biofortification traits in DH populations of rice. PLoS One. 2018;**13**:e0201756

[56] Descalsota-Empleo GI, Amparado A, Inabangan-Asilo MA, Tesoro F, Stangoulis J, Reinke R, et al. Genetic mapping of QTL for agronomic traits and grain mineral elements in rice. The Crop Journal. 2019

[57] Kumar N, Jain RK, Chowdhury VK. Linkage Mapping of QTLs for Grain Minerals (Iron and Zinc) and Physio-Morphological Traits for Development of Mineral Rich Rice (*Oryza sativa* L.) Indian Journal of Biotechnology. 2019;**18**:69-80

[58] Xu Q, Zheng TQ, Hu X, Cheng LR, Xu JL, Shi YM, et al. Examining two sets of introgression lines in rice (*Oryza sativa* L.) reveals favorable alleles that improve grain Zn and Fe concentrations. PLoS One. 2015;**10**:e0131846

[59] Friedman M, Brandon DL. Nutritional and health benefits of soy proteins. Journal of Agricultural and Food Chemistry. 2001;**49**:1069-1086

[60] Panthee DR, Pantalone VR, Saxton AM, West DR, Sams CE. Genomic regions associated with amino acid composition in soybean. Molecular Breeding. 2006;**17**:79-89

[61] D'Mello JPF. Amino acids as multifunctional molecules. In: Amino acids in animal nutrition. 2nd ed. Cambridge: CABI Publishing; 2003. pp. 1-14

[62] Hambidge M. Human zinc deficiency. The Journal of Nutrition. 2000;**130**(5):1344S-1349S

[63] FAO/WHO. Vitamin and Mineral Requirements in Human Nutrition. 2nd Ed. Geneva: World Health Organization; 2005

[64] Kühn LC. Control of cellular iron transport and storage at the molecular

level. In: Iron Nutrition in Health and Disease. London: John Libbey; 1996. pp. 17-29

[65] Harland BF, Morris ER. Phytate: A good or a bad food component. Nutrition Research. 1995;**15**:733-754

[66] Erdman JW. Bioavailability of trace minerals from cereals and legumens. Cereal Chemistry. 1981;**58**:21-26

[67] Febles CI, Arias A, Hardisson A, Rodriguez-Alvarez C, Sierra A. Phytic acid level in wheat flours. Journal of Cereal Science. 2002;**36**:19-23

[68] Liu Z, Cheng F, Zhang G. Grain phytic acid content in japonica rice as affected by cultivar and environment and its relation to protein content. Food Chemistry. 2005;**89**:49

[69] Liu QL, Xu XH, Ren XL, Fu HW, Wu DX, Shu QY. Generation and characterization of low phytic acid germplasm in rice (*Oryza sativa* L.). Theoretical and Applied Genetics. 2007;**114**:803

[70] Saini DK, Devi P, Kaushik P. Advances in genomic interventions for wheat biofortification: A review. Agronomy. 2020;**10**(1):62

[71] Kaushik P. Application of Conventional, Biotechnological and Genomics Approaches for Eggplant (*Solanum melongena* L). Breeding with a Focus on Bioactive Phenolics. March 11, 2019 [Cited: 17 July 2019]. Available from: https://riunet.upv.es/handle/10251/122295

[72] Srivastava A, Kaushik P, Singh SP. Physiological traits and their importance in wheat breeding. Trends in Biosciences. 2018;**11**(**15**):2499-2503

Chapter 5

Directing for Higher Seed Production in Vegetables

Navjot Singh Brar, Dinesh Kumar Saini, Prashant Kaushik, Jyoti Chauhan and Navish Kumar Kamboj

Abstract

Vegetables are essential for human health and well-being. For sustaining an excellent production of vegetable crops, the seed is a primary input. Moreover, good quality seed is an important requirement for the vegetable industry, and there is a huge demand that has been expanding, considering the fact that seed multiplication is economically pertinent for vegetable cultivars to contend commercially. But the healthy seed production is usually a sumptuous trait and tormented by agricultural tactics, genetics as well as by the environmental factors. Features like seed output of the vegetables, sizeable genetic variation, the prerequisite for advancement and acceptance of a good quality vegetable seed. Here different mechanisms for seed production in vegetable crops has been presented, also vital areas and factors influencing seed production, and eventually discourses regarding the opportunity of plant breeding to sustainably make improvements to vegetable seed production.

Keywords: vegetables, self-incompatibility, male sterility, seed

1. Introduction

Vegetables play a critical role in food security and are rich in mineral and vitamins. Vegetables can prevent several chronic diseases, including cancer [1]. But the successful production of vegetables depends on the first primary input that is the seed. Seeds are the consequences of sexual reproduction in the plant, and not all plants produce. Besides, the seeds are of tremendous organic and financial value. They have abundant protein, starch and oil reserves that assist during the early stages of development of a plant. These reserves are what make several portions of cereal and legumes main foods resources for any considerable proportion in the world's inhabitants [2]. Vegetable seeds signify a crucial organ intended for the multiplication of vegetables. Seeds accumulate well-balanced free of charge amino acids, which stored inside the seed storage proteins (SSPs). The seed quality is determined by the kind of amino acids, specifically crucial amino acids [3].

Vegetables require two successive processes, namely pollination and fertilisation, to produce the seeds. Pollination refers to the transfer of pollen from the androecium (male flower part) to the gynoecium (female flower part). Generally, flowers contain two other parts, the sepals and petals, which may be helpful to attract pollinators, namely insects. It is not necessary for flowers to have all of these four structures mentioned. Flowers either may be complete, having all four or incomplete, not having all four parts. Likewise, flowers may be grouped into perfect and imperfect flowers [4]. There are two main types of pollination, namely, self and cross-pollination. Self-pollination refers to the deposition of pollen from the anther on the stigma located on the same plant (geitonogamy). It is the closest form of inbreeding which leads to homozygosity. Species having this type of pollination develop homozygous balance and do not exhibit significant inbreeding depression [5].

Whereas the transfer of pollen from the anther of one flower to the stigma of another flower on a different individual is called cross-pollination. It is the form of outbreeding which leads to heterozygosity. Outbreeder species develop heterozygous balance and exhibit significant inbreeding depression on selfing. In addition to these two types of pollination, there is one a different kind of pollination, often called cross-pollination, where cross-pollination often exceeds 5% and may reach 30%. Various mechanisms such as bisexuality, homogamy, cleistogamy and position of anthers promote self-pollination, whereas other mechanisms such as dicliny (namely monoecy and dioecy), dichogamy, heterostyly, herkogamy, self-incompatibility (namely sporophytic and gametophytic) and male sterility promote cross-pollination [6].

Nearly one-third of the current global population is suffering from some form of malnutrition. Moreover, with a constant rise in the world population the food demand tends to increase up to 60% [7]. Vegetables being shorter duration crops can play a crucial role in providing more food per unit of cultivated area [8]. Similarly, climate change is a result of human activities primarily related to the emission of greenhouse gases. It means that there must be a focus on vegetable production and lowering the per capita emissions of greenhouse gases [9]. Vegetables are sensitive to temperature fluctuations, and environmental stresses have also been found to affect the nutrient composition of vegetables [10].

Monoecious, that is, cucurbits have both male and female flowers on different branches of the same plant. Dioecious (like spinach) have male (staminate) and female (pistillate) flowers on separate plants. Generally, both of these monoecious and dioecious plants require cross-pollination. Pollen grain germinates and penetrates the style to reach the ovary and then fertilises the ovule. This fertilised ovule changes into seed and the surrounding ovary develops into the fruit [11]. There are different agents for pollination such as insects, wind and water, etc. Insects are main agents for pollination in vegetables; they visit flowers to collect pollen and nectar as food and transfer this pollen unknowingly to the stigma of other flowers on the same plants or different plants. In this review, we have tried to comply with the important aspects for the successful and mass production of healthy vegetable seeds [12].

2. Pollination in vegetable crops

Crops can be classified into three categories depending upon the mode of pollination, that is, naturally self-pollinated, naturally cross-pollinated and both self and cross-pollinated crops. Naturally self-pollinated: In such plants, same floral structure or different flowers on same plant houses both pollen and embryo sac. Examples are tomato (*Solanum lycopersicum* L.), lettuce (*Lactuca sativa* L.), parsnip (*Pastinaca sativa* L.), peas (*Pisum sativum* L.), dwarf bean (*Phaseolus vulgaris* L.) (**Table 1**).

Naturally cross-pollinated: in cross-pollinated plants, male and female flowers are present on different plants. While in some cases, the stigma may not be receptive at the time of pollen availability. For example, cabbage (*Brassica oleracea var. capitata* L.), cauliflower (*Brassica oleracea var. botrytis* L.), onion (*Allium cepa* L.), broccoli (*Brassica oleracea* var. *italic* L.), carrot (*Daucas carota* L.), radish (*Raphanus sativus* L.), pumpkin (*Cucurbita moschata* Duchesne), squash (*Cucurbita pepo* L.), beet (*Beta vulgaris* L.), muskmelon (*Cucumis melo* L.), cucumber (*Cucumis sativus* L.).

Crops	Techniques used for hybrid seed production
Onion	*S and T type-cytoplasmic male sterility (CMS) with natural pollination
Carrot	Brown anther and *petaloid sterility-cytoplasmic male sterility (CMS) with natural pollination
Cole crops and radish	Oguro type-cytoplasmic male sterility (CMS) and *Sporophytic self-incompatibility (SSI) with natural pollination
Cucurbits	Genetic male sterility mainly in muskmelon and *hand emasculation with natural (pinching and use PGR for staminate flower) or hand pollination
Pepper	*Genetic male sterility with hand pollination or natural pollination
Tomato and Brinjal	*Hand emasculation and hand pollination

**Commercial use of techniques in hybrid seed production.*

Table 1.
Techniques of hybrid seed production.

Crop	Pollination type	Mechanism
Solanaceae		
Tomato (*Solanum lycopersicum*)	Self	Bisexual, stigmas surrounded by anthers
Eggplant (*S. melongena*)	Self	Bisexual, stigmas surrounded by anthers
Potato (*S. tuberosum*)	Self	Bisexual, hypogynous
Peppers (*Capsicum annuum*)	Self	Bisexual, hypogynous
Cucurbitaceae		
Bottle gourd (*Lagenaria siceraria*)	Cross	Monoecious
Watermelon (*Citrullus lanatus*)	Cross	Monoecious
Cucumber (*Cucumis sativa*)	Cross	Monoecious
Gherkin (*C. anguria*)	Cross	Monoecious
Muskmelon (*C. melo*)	Cross	Monoecious
Snake gourd (*Trichosanthes cucumerina*)	Cross	Monoecious
yellow-flowered gourd (*Cucurbita pepo ovifera*)	Cross	Monoecious
zucchini (*C. pepo*)	Cross	Monoecious
Cole crops		
Brussels sprouts (*Brassica oleracea var. gemmifera*)	Cross	Sporophytic self-incompatibility
Cabbage (*B. oleracea* var. capitata)	Cross	Sporophytic self-incompatibility
Cauliflower (*B. oleracea* var. botrytis)	Cross	Sporophytic self-incompatibility
Kale (*B. oleracea* var. sabellica)	Cross	Sporophytic self-incompatibility
Broccoli (*B. oleracea* var. italica)	Cross	Sporophytic self-incompatibility
Turnip (*B. rapa* subsp. Rapa)	Cross	Sporophytic self-incompatibility
Fabaceae		
Common Bean (*Phaseolus vulgaris*)	Self	Self-fertilisation before opening the flowers (Cleistogamous)
Faba bean (*Vicia faba*)	Partial cross pollination	Partly cleistogamous

Crop	Pollination type	Mechanism
Lima bean (*P. lunatus*)	Self	Cleistogamous flower structure
Chickpea (*Cicer arietinum* L.)	Self	Cleistogamous flower structure and stigmas surrounded by anthers
Cowpea (*Vigna unguiculata*)	Self	Cleistogamous flower structure
Fenugreek (*Trigonella foenum-graecum*)	Self	Cleistogamous flower structure
sweet pea (*Lathyrus odoratus*)	Self	Cleistogamous flower structure and stigmas surrounded by anthers
Pea (*Pisum sativum*)	Self	Cleistogamous flower structure and stigmas surrounded by anthers
Soybean (*Glycine max*)	Self	Cleistogamous flower structure and stigmas surrounded by anthers

Table 2.
Different kind of pollination mechanisms in the vegetable crops.

Other cucurbits (bitter gourd, bottle gourd, ridge gourd, sponge gourd, snake gourd, pointedgourd, ash gourd, etc.), amaranths.

Both self and cross-pollinated: Plants are primarily self-pollinated, but cross-pollination occurs to varying extents. Examples include brinjal (*Solanum melongena* L.), okra (*Abelmoschus esculentus* (L.) Moench), chilli (*Capsicum annuum* L.), sweet pepper (*Capsicum* spp.). After landing on stigma, pollen grains germinate and grow down the style of the flower, and this process is called fertilisation. Sperms of the pollen unite with ovules in the ovary which leads to seed production. In the event of pollen incompatibility, a fully pollinated flower does not get fertilised. Some plants are capable of producing fruit without fertilisation and seed production, and such species are called parthenocarpic [13, 14].

Among the different pollination agents like wind, birds, insects, gravity, water and mammals, the most important are insects. Insects contribute 80–85% of the pollination, out of which a hefty proportion of 75–80% is attributable to honey bees. Because of their body characteristics and behaviour patterns, solitary bees, bumblebees and honey bees constitute the largest group of pollinators. Pollination by insects is indispensable for improvement of plant and yield characteristics like seed set, quality of produce, early flowering, oil content, rubber content, pyrethrin content, etc. Managed pollination of crops by honey bees is a surest and most effective way of increasing yield and quality of the produce. Honey bees enhance productivity of crops through cross-pollination along with additional income through production of honey and beeswax, etc. Honey bees and other cross-pollinating agencies like bats, small mammals, birds, etc. owing to its body modifications to pick pollen, floral fidelity, efficiently communication among the colony members and their adaptability to different climates [15]. Cross-pollination results in hybrid vigour, thus improving the quality as well as quantity of the produce which is a boon for vegetable seed production (**Table 2**).

3. Effects of insect pollination on seed yield of vegetables

Inadequate pollination has been a major constraint to the potential returns of vegetables. Different insect pollinators have been identified in various vegetable crops, which increased the seed yield by increasing the pollination. Vinícius-Silva et al. [16] found fifteen floral visitors with *Exomalopsis analis* being the most representative in tomato crop. They also reported the presence of the other two effective pollinators, namely *Apis mellifera* and *Trigona spinipes* in tomato crop. Shah et al. [17] observed the highest population of honey bees among all tracked pollinators in cucumber and showed that insect pollination in cucumber acts as additional input in enhancing the yield [18].

Similarly, the highest weight of fruits, number of seeds per fruit, fruit size and TGW was achieved in honey bee pollination compared to others. Azmi et al. [19] observed heavier, longer and larger fruits in cucumber when pollinated by stingless bee (*Heterotrigona itama*) and hand compared to those produced from pollination without *H. itama*. Rouf et al. [20] reported an increase of 45.46 and 23.17% in seed yield of cauliflower plants pollinated by honey bee over plants grown inside net without bees and open pollination, respectively. Further, they showed that maximum yield attributes of seed could be achieved if planned bee pollination and central curd cutting employed together.

4. Plant ideotype for seed production

For the first time in 1968, Donald introduced the concept of ideotype in plant breeding. Later in 1976, the concepts of isolation, competition and crop ideotypes were proposed by Donald and Hamblin [21].

4.1 Ideotypes for solanaceous vegetables

Manipulation of plant architecture of tomato may provide increased fruit yield resulting in increased seed yield. Suarma et al. [22] suggested emphasising on traits such as fruit yield (q/ha), plant height, average fruit weight for ideotype construction in tomato. Direct selection for these traits, having high heritability and genetic advance, may yield expected genetic up-gradation of a genotype. Sarlikioti et al. [23] suggested a new plant ideotype for optimization of light absorption and canopy photosynthesis in tomato. This new ideotype with more spacious canopy architecture due to long internodes and long and narrow leaves led to an increase in crop photosynthesis of up to 10%. Recently, Zsögön et al. [24] suggested that vital monogenic traits whose physiology has been revealed thoroughly can be molecularly tailored using genome editing techniques to achieve the target ideotype for elite cultivars of tomato. They also proposed that wild relatives or progenitors harbouring polygenic traits of interest could be de novo domesticated by manipulating monogenic yield-related characters through these techniques to get 'model type' plants which would perform expectedly in a defined environment. It has been suggested that shifting of crop plants from annuals to perennials may provide an additional advantage in seed yield. Eggplant ideotypes characterised by a radical change in plant architecture, with an arborescent or shrubby habit and perennial instead of annual fruit set using somatic hybridization [25–27].

4.2 Ideotypes for cucurbits

Plant architecture of muskmelon has also been manipulated to get increased fruit yield. Two different plant ideotypes have been proposed to get increase fruit set in muskmelon: "bush" or "birdnest" type possessing multilateral branches of the same length and bearing uniform sized fruits near the centre of plants and short internodes types having indeterminate growth behaviour and shorter internodes which can be planted at higher densities [28].

4.3 Ideotypes for fabaceous vegetables

Manipulation of the architecture of plants to achieve high seed production has been accomplished in various fabaceous species such as common bean, broad bean and pea; and also in the underexploited species of this family [29, 30].

Isaacs et al. [31] employed participatory plant breeding approach and together with farmers, identified specific traits that constitute a bean ideotype: adaptation, restricted height, columnar plant structure, even distribution of pods, fewer leaves, and earlier maturity. Plants with this ideotype produced good seed yield and were suitable for maize-bean cropping systems. Polania et al. [32] 2017 evaluated 36 bean genotypes to test the relationships between shoot traits and root traits under drought conditions. They identified two ideotypes related to efficient water use: water savers having a shallower root system and water spenders presenting more in-depth root system. Both showed greater root vigour under drought stress and produced high grain yield. Recently, Bodner et al. [33] identified ideotypes, having higher average yield, taller structure, more pods per node and longer flowering duration, suitable for Northern Europe. They considered Baltic landraces as promising ideotypes for increased *V. faba* yields in Nordic target environments as well as the other workers [34].

5. Seed set and development

Since all vegetables are angiosperms, so a standard procedure of fertilisation, seed set and development is followed in all vegetables with few modifications. We are presenting here a general mechanism of fertilisation, seed set and development. At the time of fertilisation, protective coats, known as integuments and a central tissue called nucellus are present in the angiosperm ovule. If we see the structure of ovule, clear differentiation of these two integuments and nucellus can be found in the region of the micropyle, it is a minute pore in the integuments through which, the pollen tube enters the nucellus and move towards egg cell and polar nuclei. A stalk, funicle, attaches the ovule to the wall of the ovary. In general, megaspore mother cell inside the nucellus once divides meiotically and then divides mitotically three times to produce embryo sac or female gametophyte, a haploid eight-nucleate, seven-celled structure which comprises of one egg cell, two synergids, three antipodal and two polar nuclei. Although among angiosperms, the female gametophyte has a variety of forms, it may not necessarily encompass all these seven cells. On the other hand, inside the anther, microspore mother cell first divides meiotically and then mitotically to produce pollen grain or microgametophyte, which comprises two sperm cells enclosed with one vegetative cell [35].

These two female and male gametophytes play essential roles in the reproductive process of angiosperm. Sexual reproduction starts with the transmission of male gametophyte or pollen grain from anther to the carpel's stigma. Subsequently, pollen grain begins to germinate on stigma and a pollen tube carrying two sperm cells is formed, which penetrates the style. Growth and development of pollen tube is controlled by vegetative nucleus which disintegrates after serving its duty. Pollen tube enters into the embryo sac through micropyle, in general, and releases two male gametes. One male gamete fertilises the egg cell, called syngamy, and the second male gamete fuses with the central cell or polar nuclei [36]. Since two successive fertilisations take place, the procedure is known as double fertilisation. The zygote is formed after uniting of one sperm cell with egg cell, and this zygote gives rise to seed's embryo which is the starting of the sporophyte generation. Following fertilisation, central cell's polar nuclei produce seed's endosperm, which is the nutrition source for developing embryo. These two embryos and endosperm encompass the central portion of the seed. Two synergids and three antipodal, remaining five nuclei, do not play any further role in seed development. For the development of viable seed, successful fertilisation of egg cell and the central cell is necessary [37]. All seeds mostly contain an embryo, a protective cover-seed coat and a reserve of food materials or any other specified tissue such as perisperm. Occasionally, polyembryony condition refers to development of more than one embryo in a single seed, may also be observed in some families such as Solanaceae and Amaryllidaceae families.

6. Advances in the understanding of molecular mechanisms underlying seed/fruit set and development

Underlying molecular mechanisms of seed set and development in angiosperms is becoming clear rapidly with the advancements of various omics studies such as genomics, transcriptomics and proteomics etc. and genetic transformation techniques. These mechanisms are generally conserved across all the angiosperms and may also be operated in vegetables. Various studies have been conducted in vegetables such as tomato (and cucumber to explore the underlying molecular mechanism of seed set and fruit set) [38]. In 2016, isolated and characterised two allelic mutants, twisted seed1-1 (tws1-1) and tws1-2 of a single copy gene (TWS1). This gene encodes a small protein of 81 amino acids which regulates embryonic development and accumulation of storage compounds in the seed [39].

This gene is specifically conserved among angiosperms and can be cloned from vegetables to explore its function in seed development in vegetables. The importance of AN3-MINI3 gene cascade in seed embryo development. Their regulatory model provided a deep insight into the seed mass regulation, which may be further explored to increase seed yields of vegetables [40]. Role of mitochondrial reactive oxygen species homeostasis in gametophyte and seed development has also been highlighted in angiosperms. It was reported that the effect of the mutation in *AtHEMN1* gene which encodes for coproporphyrinogen III oxidase. They showed adverse effects of *Athemn1* mutant alleles on gametophytic and seed development. Adverse effects included the development of nonviable pollen and embryo sacs with unfused polar nuclei, defects in endosperm development due to abnormal differentiation of the central cell and arresting of embryo development at the globular stage [41].

To ensure successful sexual plant reproduction, fruit set or transformation of flowers to fruits is very critical. Role of hormones (i.e. auxin and gibberellins) in controlling fruit set after pollination and fertilisation have been well understood. It was shown that the role of microRNA-based (microRNA159/GAMYB1 and −2 pathway) regulation ovary development and fruit set in tomato. They initiated fruit set by modulating auxin and gibberellin responses using SlGAMYBs. On the other hand, proteins such as TIR1-like proteins have also been shown to have essential roles in auxin-mediated fruit development processes. Two TIR1-like genes have been identified in cucumber and designated as CsTIR1 and CsAFB2 [42]. Xu et al. [43] used tomato as a model plant to investigate the effects of these two genes on fruit/seed set. They highlighted the crucial role of the miR393/TIR1 component in fruit/seed set and concluded that post-transcriptional regulation of these two genes mediated by miR393 is vital for fruit set initiation in both cucumber and tomato. The different stages of seed development and the structure of a dicot seed is presented in **Figure 1**.

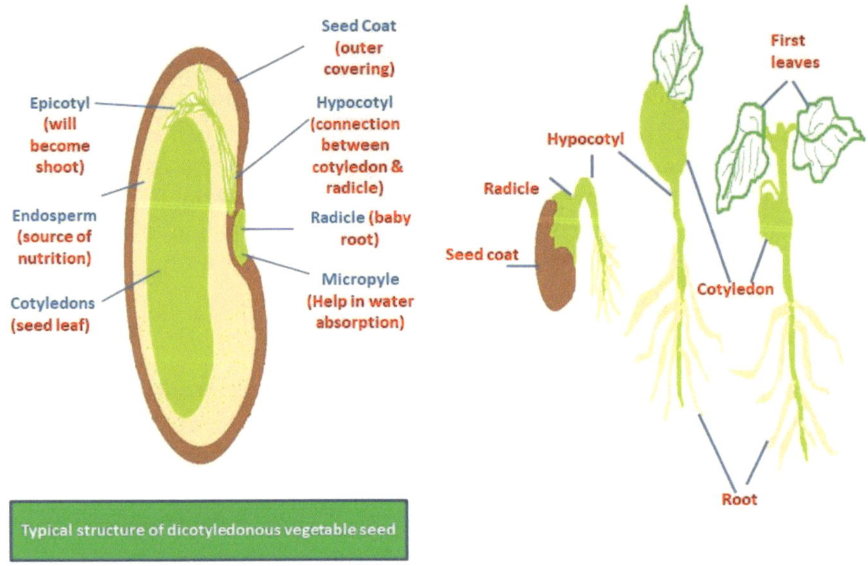

Figure 1.
Different stages of seed development.

7. Role of male sterility in vegetables

7.1 Family: solanaceae

7.1.1 Pepper/Chilli (Capsicum annuum L.)

Peterson (1958) first reported the cytoplasmic genetic male sterility (CGMS) in chilli in an introduction of *C. annuum* from India (PI-164835) and found its instability under fluctuating conditions, particularly temperatures and natural cross-pollination. Genetic male sterility in chilli well exploited on a commercial scale in hybrid seed production. Male sterile plants easy to identify in the field at a comparatively early stage. Nearly 20 genes governed genetic male sterility. The ms-10 gene is linked with taller plant height, erect growth and dark purple anther. MS-12 (ms-509/ms-10) and ms-3 genes are commercially utilised in India and Hungry, respectively [44].

7.1.2 Tomato (Solanum lycopersicum L.) *and brinjal* (Solanum melongena L.)

Tomato crop has different types of male sterility identified, but presently commercial hybrid seed production in tomato and brinjal possible with manual emasculation and hand pollination and it is economically viable and dominates in the seed industry. Though, the availability of different sterility methods can be used to avoid selfing and optimise crossing resulted in reducing the cost of hybrid seed production [45].

7.2 Cucurbitaceous vegetables

To exploit heterosis in cucurbits the essential requirement is heterotic combination potential of crops from flower size to pollination and fruit set resulted in proper seed setting to economic feasibility. The cucurbit vegetables have a more substantial size of male and female flowers and allow following other systems of pollination control strategies. The hand emasculation with hand and natural pollination mechanism used in hybrid seed production in bottle gourd, pumpkin, squash, cucumber, muskmelon and bitter gourd with specific planting ratio. Genetic male sterility mainly uses in muskmelon, and most of the genetic male-sterile mutants in cucurbits are monogenic recessive. There are many types of male sterility identified in cucurbits, but commercial exploitation is still lacking. Gynoceious lines based on genetic male sterility (GMS) stability gene and use of different plant growth regulators are also useful in hybrid seed production of cucurbits with sex modification [46].

7.3 Cole crops (*Brassica oleracea* L.) and radish (*Raphanus sativus*)

Cole crops and some root crops are a significant group of vegetables in the brassica family, and they are cabbage, cauliflower, broccoli, turnip and radish. GMS, CGMS and self-incompatibility (SI) are important pollination mechanism available in Brassica family to get a higher percent of heterosis in crops. In which, self-incompatibility (Sporophytic self-incompatibility) system is most useful in hybridisation program. But, CGMS method also developed with some self-pollination occurrence [47]. In cole vegetables, sterile cytoplasm (CMS system) derived from *B. nigra* through interspecific hybridisation between *B. nigra* and *B. oleracea* var. italica and Ogura type CMS also identified and reported in cultivar Japanese radish of *Raphanus sativus*. First, introgression of this sterility cytoplasm to *Brassica oleraceae* genome through repeated backcrosses with broccoli. Some plant physiological problems were found in Ogura based CMS lines of broccoli, cauliflower, cabbage, Brussels sprout and it has been solved using protoplast fusion, and this technique is also used in transfer Ogura cytoplasm from broccoli into cabbage [48].

7.4 Carrot (*Daucus carota* L.)

Cytoplasmic male sterility in carrot can occur in two morphologically (brown anther and petaloid) distinct phenotypes. The brown anther male sterility was first discovered in the cultivar Tendersweet, and this is, characterised by shrivelled, yellow-to-brown anthers with no pollen. It is a homeotic mutation. This is established as the white petaloidy or green petaloidy. It is stable male sterility across a wide range of environments as compared to brown anther type. Seed yield of the brown-anther CMS are generally higher because of petaloid sterility shows less frequent deterioration to male fertility [49].

7.5 Bulb crops—onion (*Allium cepa* L.)

Male sterility in onion, first reported in the progenies of an onion cultivar Italian Red plants and is controlled by the combination of a cytoplasmic factor "S" together with a recessive nuclear restorer locus in its homozygous form (ms) and "T"-cytoplasm has been reported. Onion (*Allium cepa* L.) hybrid seed production has been produce in all over the world through CGMS-based systems in which mostly hybrids are derived from S-cytoplasm because of its stability in various environments [50, 51].

8. Self-incompatibility

Self-incompatibility can be a widespread phenomenon in vegetable crops that forestalls inbreeding and encourages outcrossing. The response of self-incompatibility is genetically managed by several multi-allelic loci and depends on many intricate interactions among the self-incompatible pollen and pistil combinations. It is genetically regulated phenomena that function as a barrier to self-pollination in the big selection of vegetable crops like cabbage, cauliflower, tomato and many others. Self-incompatibility can be a critical system by which crops avert self-fertilisation and keep a broad genetic range. Self-incompatibility is considered to present in 30–50% of flowering plant species [52]. Many SI programs have now been discovered. In all situations, incompatible (self-) pollen is considered by a distinct system usually genetically managed that brings about inhibition on the pollen while in the stigma or on the pistil. Using SI in F1 hybrid generation has key gain over other approaches. Usage of Self-incompatibility in cole crops for hybrid seed generation is commercialised due to the availability of a robust mechanism/method to create large-scale F1 seeds employing picked parental strains is undoubtedly a critical issue, which in the long run establishes the professional viability on the hybrid varieties [53].

Self-incompatibility is classified as namely gametophytic and sporophytic. In gametophytic technique self-incompatibility response of pollen and stigma is decided with the genotype of the female plant on which pollens are developed (e.g. tomato) even though in sporophytic technique, pollen phenotype (self-incompatibility response) is identified with the genotype on the female plant on which pollens are developed (e.g. cole greens). In Brassicaceae, sporophytic self-incompatibility (SSI) has been ideally characterised and productively used for that growth of commercial hybrids. Using SI in F1 hybrid generation has key gain over other approaches; equivalent portions of seed on the two inbred strains can be blended jointly for demonstrating, along with the total crop is harvested for seed. For hybrid seed generation, equally the parental inbreds need to have two diverse S alleles for sturdy self-incompatibility in the event of one cross hybrid. Among the cole greens like cabbage, cauliflower, broccoli and many others, sporophytic self-incompatibility system is currently being used for the hybrid seed generation at many spots in India [54]. Usually in cauliflower self-incompatibility is weak, and its response is broken at substantial temperature. Self-incompatibility can be a technique employed by a lot of flowering plant species to forestall self-fertilisation and thus encourage outcrossing. Above the several years, considerable perception in the mechanisms regulating self-incompatibility has become attained for that Solanaceae gametophytic self-incompatibility programs at the same time as for that sporophytic self-incompatibility technique of the Brassicaceae in vegetable crops. A mix of genetic and molecular reports have resulted while in the identification and characterisation of the self-incompatibility genes associated with this particular reaction.

Moreover, careful investigation on the factors in the signalling cascades of equally the Solanaceae along with the Brassicaceae is necessary for an entire idea of the self-incompatibility reaction in these people. Several mechanisms and approaches have not been exploited for that growth of professional hybrids in vegetable crops between that SI is of crucial relevance. While in the light-weight of the quick progression of biotechnology, it could be expected that SI programs are going to be ever more used near foreseeable future, in vegetable crops [55].

9. Growth regulators

Growth regulators are organic chemical substances which, when applies in small quantities aid in the regulation of plant growth and modify the physiological response in plants. Growth regulators have immense importance in enhancing vegetable production and have been used to improve seed germination, increase in yield and tolerance against diseases and unfavourable conditions [56]. Apart from these functions growth regulators have usefulness in vegetable seed production by altering sex expression, increasing fruit set as well as seed yield and inducing male sterility, without exerting any harmful effects on the environment and human health [57]. Classification and functions of different plant growth regulators (PGRs) are listed below:

1. Auxins (IAA, NAA, IBA, 2,4-D, 4-CPA): apical dominance, root induction, control fruits drops, regulation of flowering.

2. Gibberellins (GA3): seed germination, stimulates flowering, increase flower and fruit size.

3. Cytokinins (kinetin, zeatin): bud initiation and root growth, storage life prolongation of vegetables.

4. Ethylene (etheral): uniform ripening in vegetables, promotes abscission, senescence of leaf.

5. Abscisic acid (dormins, phaseic acid): stress hormone, dormancy, seed development and germination.

6. Flowering hormones (florigen, vernalin).

7. Natural substances (vitamins, phytochrome tranmatic).

8. Synthetic substances (synthetic auxins, synthetic cytokinins).

Role of different PGRs in vegetable production of different vegetable crops are reviewed in **Table 3**:

PGR	Target/response	Crop
GA3	Fruit setting, seed yield and quality	Bittergourd, muskmelon, tomato, chilli, capsicum, brinjal, cauliflower, cabbage, okra, cucurbits, potato, pea
GA3	Abnormalities in pollen development and induced the carpelization of stamens	Pepper
GA3	Leaf morphogenesis, promote normal stamen and pollen development	Tomato
GA3	Production of male sterile flowers	Onion, Brussels sprouts, cabbage, cauliflower and kale
GA3	Increased number of female flowers	Bitter gourd
GA3	Lower male and female flower ratio	Cucumber
GA3	Induce parthenocarpy	Bitter gourd
Ethrel	Decreased number of staminate flowers	Cucumber, bittergourd, pumpkin, sponge gourd
Ethrel	Increased number of pistillate flowers	Cucumber, pumpkin, pointed gourd, melons, snake gourd, sponge gourd, bottle gourd, bitter gourd, summer squash

PGR	Target/response	Crop
Ethrel	Lower male female flower ratio	Cucumber
Ethrel	Increased yield	Cucumber, bitter gourd, cucurbits, potato, pumpkin
Ethrel	Induction of male sterility	Lettuce, eggplant, squash
TIBA (triiodobenzoic acid)	Induction of male sterility	Tomato
TIBA (triiodobenzoic acid)	Producing a favourable female to male ratio and increased number of fruits	Cucumber, squash, watermelon
MH (maleic hydrazide)	Induction of male sterility	Tomato, coriander, pepper, okra, onion, squash, chilli, eggplant
MH (maleic hydrazide)	Decreased number of male flowers	Cucumber, sponge gourd
NAA (naphthalene acetic acid)	Induce male sterility	Tomato and squash
NAA (naphthalene acetic acid)	Reduce staminate-pistillate flower ratio	Cucumber, squash
NAA (naphthalene acetic acid)	Increased number of female flowers	Cucurbits, sponge gourd
NAA (naphthalene acetic acid)	Decreased number of male flowers	Cucumber
NAA (naphthalene acetic acid)	Induce parthenocarpy	Bitter gourd
NAA (naphthalene acetic acid)	Increased fruit set/yield	Cucumber, bottle gourd, tomato, chilli, capsicum, brinjal, cauliflower, cabbage, onion, garlic, cucurbits, okra, tomato
Dalapon (dichloropropionic acid)	Induction of male sterility	Pea, tomato
Dalapon and a-chloropropionate	Supression of anther dehiscence	Pepper tomato
FW-450 (sodium 2,3-dichloroisobutyrate)	Induction of male sterility	Tomato
CCC ((2-chloroethyl) trimethylammonium chloride)	Selectively inhibited the development of stamen or suppressed pollen	Tomato
ABA (abscisic acid)	Selectively inhibited the development of stamen or suppressed pollen	Tomato
Indole acetic acid (IAA)	Increased pistillate flowers	Cucurbits, cucumber
Indole acetic acid (IAA)	Decreased male flowers	Cucumber
Indole acetic acid (IAA)	Improved yield and quality characteristics	Okra, cauliflower

Table 3.
Growth regulators used for higher seed production in the vegetables based on Prajapati et al. [58].

10. Conclusions

The essence of any seed programme is the excellent quality of seed, and this trait varies from the standpoint of genetic purity. The seed programme with no proper quality management of the seed will tend to fail. For that reason, the quality of vegetable seed is a necessary consideration. Underneath a standard seed technology chain, breeder seed is multiplied from nucleolus seed. The exercise of bulk enhance of breeder seed and endless multiplication cycles of basis seed with no likely again to breeder seed may severely influence the standard of seed and may be discontinued. Importance of good quality seed can be determined from the fact that seed is the indispensable input for crop production. The top-quality seed is the carrier of the resistance gene or good genes selected by the breeder. Seed ensures food supply under adverse production sites; therefore, the importance of seeds for vegetable production cannot be denied.

Conflict of interest

The authors declare no conflict of interest.

Author details

Navjot Singh Brar[1], Dinesh Kumar Saini[2], Prashant Kaushik[3*], Jyoti Chauhan[4] and Navish Kumar Kamboj[5]

1 Department of Vegetable Science, Punjab Agricultural University, Ludhiana, India

2 Department of Plant Breeding, Punjab Agricultural University, Ludhiana, India

3 Instituto de Conservación y Mejora de la Agrodiversidad Valenciana, Universitat Politècnica de València, Valencia, Spain

4 Department of Botany, Kurukshetra University, Kurukshetra, Haryana, India

5 Department of Agronomy, CCS Haryana Agricultural University, Hisar, India

*Address all correspondence to: prakau@doctor.upv.es

References

[1] Kaushik P, Andújar I, Vilanova S, Plazas M, Gramazio P, Herraiz FJ, et al. Breeding vegetables with increased content in bioactive phenolic acids. Molecules. 2015;**20**(10):18464-18481

[2] Copeland LO, McDonald MB. Seed formation and development. In: Copeland LO, McDonald MB, editors. Principles of Seed Science and Technology [Internet]. Boston, MA: Springer US; 1999. pp. 17-39. DOI: 10.1007/978-1-4615-1783-2_2

[3] Amir R, Galili G, Cohen H. The metabolic roles of free amino acids during seed development. Plant Science. 2018;**275**:11-18

[4] Rashid MA, Singh DP. A manual on vegetable seed production in Bangladesh. In: AVRDC-USAID-Bangladesh Project. Bangladesh: Horticulture Research Centre; 2000

[5] Faegri K, Van Der Pijl L. Principles of Pollination Ecology. Elsevier; 2013

[6] Barmentlo SH, Meirmans PG, Luijten SH, Triest L, Oostermeijer JGB. Outbreeding depression and breeding system evolution in small, remnant populations of Primula vulgaris: Consequences for genetic rescue. Conservation Genetics. 2018;**19**(3):545-554

[7] Islam SMF, Karim Z. World's demand for food and water: The consequences of climate change. Desalination [Internet]. 2019. Available from: https://www.intechopen.com/online-first/world-s-demand-for-food-and-water-the-consequences-of-climate-change

[8] Eigenbrod C, Gruda N. Urban vegetable for food security in cities. A review. Agronomy for Sustainable Development. 2015;**35**(2):483-498

[9] Ritchie H, Roser M. CO_2 and greenhouse gas emissions. Our World in Data. 2017. Available from: https://ourworldindata.org/co2-and-other-greenhouse-gas-emissions

[10] Singh H, Sethi S, Kaushik P, Fulford A. Grafting vegetables for mitigating environmental stresses under climate change: A review. Journal of Water and Climate Change. 2019

[11] Dellaporta SL, Calderon-Urrea A. Sex determination in flowering plants. The Plant Cell. 1993;**5**(10):1241-1251

[12] Willmer P. Pollination and Floral Ecology. Woodstock, England: Princeton University Press; 2011

[13] Singh P. Manual on Vegetable Breeding. New Delhi, India: Daya Publishing House. 2014. pp. 29-57

[14] Rana MK, Kaushik P. Pigeon pea. In: Vegetable Crop Science. Boca Raton, Florida: CRC Press; 2017. pp. 715-722

[15] Colombo N, Galmarini CR. The use of genetic, manual and chemical methods to control pollination in vegetable hybrid seed production: A review. Plant Breeding. 2017;**136**(3):287-299

[16] Vinícius-Silva R, Parma D d F, Tostes RB, Arruda VM, Werneck M d V, Vinícius-Silva R, et al. Importance of bees in pollination of *Solanum lycopersicum* L. (Solanaceae) in open-field of the Southeast of Minas Gerais State, Brazil. Hoehnea. 2017;**44**(3):349-360

[17] Shah I, Shah M, Khan A, Usman A. Response of insect pollinators to different cucumber, *Cucumis sativus* L. (Cucurbitales: Cucurbitaceae) varieties and their impact on yield. Journal of Entomology and Zoology Studies. 2015;**3**(5):374-378

[18] Thakur M, Rana RS. Studies on the role of insect pollination on cucumber yield. Pest Technology. 2008;**2**(2):130-133

[19] Azmi W, Samsuri N, Firdaus M, Hatta M, Ghazi R, Chuah T-S. Effects of stingless bee (*Heterotrigona itama*) pollination on greenhouse cucumber (*Cucumis sativus*). Malaysian Applied Biology. 2017;**46**:51-55

[20] Rouf M, Rahim M, Siddique M, Meah M. Effect of honey bee pollination and curd scooping on seed yield of cauliflower. Bangladesh Journal of Agricultural Research. 2016;**41**:251

[21] Zhu L, Zhang D-Y. Donald's ideotype and growth redundancy: A pot experimental test using an old and a modern spring wheat cultivar. PLoS One. 2013;**8**(7):e70006

[22] Suarma JP, Singh AK, Kumar S, Kumar S. Identification of traits for ideotype selection in tomato (*Lycopersicon esculentum* Mill.). Mysore Journal of Agricultural Sciences. 2009;**43**(2):222-226

[23] Sarlikioti V, de Visser PHB, Buck-Sorlin GH, Marcelis LFM. How plant architecture affects light absorption and photosynthesis in tomato: Towards an ideotype for plant architecture using a functional-structural plant model. Annals of Botany. 2011;**108**(6):1065-1073

[24] Zsögön A, Cermak T, Voytas D, Peres LEP. Genome editing as a tool to achieve the crop ideotype and de novo domestication of wild relatives: Case study in tomato. Plant Science. 2017;**256**:120-130

[25] Kaushik P. Application of conventional, biotechnological and genomics approaches for eggplant (*Solanum melongena* L). Breeding with a Focus on Bioactive Phenolics. Valencia, Spain: Universitat Politecnica de Valencia; 2019. pp. 78-96. Available from: https://riunet.upv.es/handle/10251/122295

[26] Kaushik P. Genetic analysis for fruit phenolics content, flesh color, and browning related traits in eggplant (*Solanum melongena*). International Journal of Molecular Sciences. 2019;**20**(12):2990

[27] Saini DK, Kaushik P. Visiting eggplant from a biotechnological perspective: A review. Scientia Horticulturae. 2019;**253**:327-340

[28] Staub JE, Zalapa JE, Paris MK, McCreight JD. Selection for lateral branch number in melon (*Cucumis melo*). In: The 8th Eucarpia Meeting on Cucurbit Genetic and Breeding, Olomuc, Czech Republic. 2004. pp. 381-388

[29] Kelly JD, Cichy KA, Siddiq M, Uebersax MA. Dry bean breeding and production technologies. In: Dry Beans and Pulses: Production, Processing, and Nutrition. Chichester, UK: Wiley-Blackwell; 2013. pp. 23-54

[30] Kaushik P, Dhaliwal MS, Kumar S. Usefulness of three under exploited vegetable legumes—A review. International Journal of Current Microbiology and Applied Sciences. 2018;**7**:861-870

[31] Isaacs KB, Snapp SS, Kelly JD, Chung KR. Farmer knowledge identifies a competitive bean ideotype for maize–bean intercrop systems in Rwanda. Agriculture & Food Security. 2016;**5**(1):15

[32] Polania J, Poschenrieder C, Rao I, Beebe S. Root traits and their potential links to plant ideotypes to improve drought resistance in common bean. Theoretical and Experimental Plant Physiology. 2017;**29**(3):143-154

[33] Bodner G, Kronberga A, Lepse L, Olle M, Vågen I, Rabante L, et al. Trait

identification of faba bean ideotypes for northern European environments. European Journal of Agronomy. 2018;**96**:1-12

[34] Kumar P, Bishnoi S, Kaushik P. Genetic variability, heritability and genetic advance for seed yield and other agro-morphological traits in fababean (*Vicia faba* L.) genotypes of different origin. Trends in Biosciences. 2018;**10**(4):28

[35] Drews GN, Koltunow AMG. The Female Gametophyte. Arabidopsis Book [Internet]. 2011;**9**:1-24. Available from: https://www.ncbi.nlm.nih.gov/pmc/articles/PMC3268550/

[36] Sundaresan V, Alandete-Saez M. Pattern formation in miniature: The female gametophyte of flowering plants. Development. 2010;**137**(2):179-189

[37] Richards AJ. Plant breeding systems. Garland Science. London, United Kingdom: Chapman and Hall; 1997. pp. 113-119

[38] Bonghi C, Manganaris GA. Systems Biology Approaches Reveal New Insights on the Molecular Regulatory Mechanism Responsible for Global Quality of Flesh Fruit. Boca Raton, Florida: CRC Press; 2012

[39] Fiume E, Guyon V, Remoué C, Magnani E, Miquel M, Grain D, et al. TWS1, a novel small protein, regulates various aspects of seed and plant development. Plant Physiology. 2016;**172**(3):1732-1745

[40] Meng L-S, Wang Y-B, Loake GJ, Jiang J-H. Seed embryo development is regulated via an AN3-MINI3 gene cascade. Frontiers in Plant Science. 2016;7:1645

[41] Pratibha P, Singh SK, Srinivasan R, Bhat SR, Sreenivasulu Y. Gametophyte development needs mitochondrial coproporphyrinogen III oxidase function. Plant Physiology. 2017;**174**(1):258-275

[42] Daniell H. Molecular strategies for gene containment in transgenic crops. Nature Biotechnology. 2002;**20**(6):581

[43] Xu J, Li J, Cui L, Zhang T, Wu Z, Zhu P-Y, et al. New insights into the roles of cucumber TIR1 homologs and miR393 in regulating fruit/seed set development and leaf morphogenesis. BMC Plant Biology. 2017;**17**(1):130

[44] Dhaliwal MS, Jindal SK. Induction and exploitation of nuclear and cytoplasmic male sterility in pepper (Capsicum spp.): A review. The Journal of Horticultural Science and Biotechnology. 2014;**89**(5):471-479

[45] Kaushik P, Dhaliwal MS. Diallel analysis for morphological and biochemical traits in tomato cultivated under the influence of tomato leaf curl virus. Agronomy. 2018;**8**(8):153

[46] Kumar A, Dixit S, Ram T, Yadaw RB, Mishra KK, Mandal NP. Breeding high-yielding drought-tolerant rice: Genetic variations and conventional and molecular approaches. Journal of Experimental Botany. 2014;**65**(21):6265-6278

[47] Mishra S, Kumari V. A review on male sterility-concepts and utilization in vegetable crops. International Journal of Current Microbiology and Applied Sciences. 2018;7(2):3016-3034

[48] Renard M, Delourme R, Mesquida J, Pelletier G, Primard C, Boulidard L, et al. Male sterilities and F1 hybrids in Brassica. In: Reproductive Biology and Plant Breeding. Manhattan, New York City: Springer; 1992. pp. 107-119

[49] Yamagishi H, Bhat SR. Cytoplasmic male sterility in Brassicaceae crops. Breeding Science. 2014;**64**(1):38-47

[50] Dhall RK. Status of male sterility in vegetables for hybrid development. A review. Advances in Horticultural Science. 2010;**24**(4):263-279

[51] Brar NS, Kaushik P, Dudi BS. Assessment of natural ageing related physio-biochemical changes in onion seed. Agriculture. 2019;**9**(8):163

[52] Silva NF, Goring DR. Mechanisms of self-incompatibility in flowering plants. Cellular and Molecular Life Sciences. 2001;**58**(14):1988-2007

[53] Nasrallah JB, Liu P, Sherman-Broyles S, Schmidt R, Nasrallah ME. Epigenetic mechanisms for breakdown of self-incompatibility in interspecific hybrids. Genetics. 2007;**175**(4):1965-1973

[54] Hiscock SJ. Pollen recognition during the self-incompatibility response in plants. Genome Biology. 2002;**3**(2):reviews1004.1-reviews1004.6

[55] Narbona E, Ortiz PL, Arista M. Linking self-incompatibility, dichogamy, and flowering synchrony in two euphorbia species: Alternative mechanisms for avoiding self-fertilization? PLoS One. 2011;**6**(6):e20668

[56] Fahad S, Hussain S, Saud S, Hassan S, Ihsan Z, Shah AN, et al. Exogenously applied plant growth regulators enhance the morpho-physiological growth and yield of rice under high temperature. Frontiers in Plant Science. 2016;7:1-13. Available from: https://www.ncbi.nlm.nih.gov/pmc/articles/PMC5003834/

[57] Phetpradap S, Hampton JG, Hill MJ. Effect of hand pinching and plant growth regulators on seed production of field grown hybrid dahlia. New Zealand Journal of Crop and Horticultural Science. 1994;**22**(3):313-320

[58] Prajapati S, Jamkar T, Singh O, Raypuriya N, Mandloi R, Jain P. Plant growth regulators in vegetable production: An overview. Plant Archives. 2015;**15**:619-626

Chapter 6

Unmanned Ground Vehicles for Smart Farms

Pablo Gonzalez-De-Santos, Roemi Fernández, Delia Sepúlveda, Eduardo Navas and Manuel Armada

Abstract

Forecasts of world population increases in the coming decades demand new production processes that are more efficient, safer, and less destructive to the environment. Industries are working to fulfill this mission by developing the smart factory concept. The agriculture world should follow industry leadership and develop approaches to implement the smart farm concept. One of the most vital elements that must be configured to meet the requirements of the new smart farms is the unmanned ground vehicles (UGV). Thus, this chapter focuses on the characteristics that the UGVs must have to function efficiently in this type of future farm. Two main approaches are discussed: automating conventional vehicles and developing specifically designed mobile platforms. The latter includes both wheeled and wheel-legged robots and an analysis of their adaptability to terrain and crops.

Keywords: smart farm, precision agriculture, agricultural robot, unmanned ground vehicle, autonomous robot

1. Introduction

The world's human population increases by approximately 240,000 people every day: it is expected to reach 8 billion by 2025 and approximately 9.6 billion by 2050. Cultivated land is at a near-maximum, yet estimates predict that food production must be increased by 70% for worldwide peace to persist circa 2050 [1]. Thus, producing sufficient food to meet the ever-growing demand for this rising population is an exceptional challenge to humanity. To succeed at this vital objective, we must build more efficient—yet sustainable—food production devices, farms, and infrastructures. To accomplish that objective, the precision farming concept—a set of methods and techniques to accurately manage variations in the field to increase crop productivity, business profitability, and ecosystem sustainability—has provided some remarkable solutions.

Figure 1 summarizes the cycle of precision agriculture and distinguishes the activities based on analysis and planning (right) from those that rely on providing motion (left). The solutions for activities illustrated in **Figure 1** right are being based on information and communication technologies (ICT), whereas the activities on the left rely on tractors, essential devices in current agriculture, that are being automated and robotized and will be also critical in future agriculture (smart farms).

The activities indicated in **Figure 1** left can be applied autonomously in an isolated manner, i.e., a fertilization-spreading task, can be performed autonomously

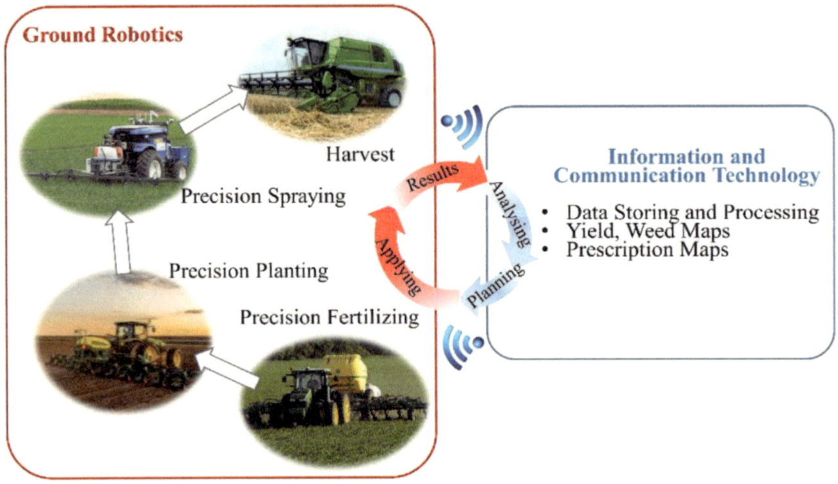

Figure 1.
UGVs in the cycle of precision agriculture.

if the appropriate implement tank has been filled with fertilizer and attached to a fueled autonomous tractor (UGV); the same concept is applicable to planting and spraying. In addition, harvesting systems must offload the yield every time their collectors are full. However, tasks such as refilling, refueling/recharging, implement attachment, and crop offloading are currently primarily performed manually. The question that arises is: would it be possible to automate all these activities? And if so, would it be possible to combine these activities with other already automated farm management activities to configure a fully automated system resembling the paradigm of the fully automated factory? Then, the combination becomes a fully automated farm in which humans are relegated to mere supervisors. Furthermore, exploiting this parallelism, can we push new developments for farms to mimic the smart factory model? This is the smart farm concept that represents a step forward from the automated farm into a fully connected and flexible system capable of (i) optimizing system performances across a wider network, (ii) learning from new conditions in real- or quasi-real time, (iii) adapting the system to new conditions, and (iv) executing complete production processes in an autonomous way [2]. A smart farm should rely on autonomous decision-making to (i) ensure asset efficiency, (ii) obtain better product quality, (iii) reduce costs, (iv) improve product safety and environmental sustainability, (v) reduce delivery time to consumers, and (vi) increase market share and profitability and stabilize the labor force.

Achieving the smart farm is a long-term mission that will demand design modifications and further improvements on systems and components of very dissimilar natures that are currently being used in agriculture. Some of these systems are outdoor autonomous vehicles or (more accurately) UGVs, which are essential in future agriculture for moving sensors and implementing to cover crop fields accurately and guarantee accurate perception and actuation (soil preparation, crop treatments, harvest, etc.). Thus, this chapter is devoted to bringing forward the features that UGVs should offer to achieve the smart farm concept. Solutions are focused on incorporating the new paradigms defined for smart factories while providing full mobility of the UGVs. These two activities will enable the definition of UGV requirements for smart farm applications.

To this end, the next section addresses the needs of UGVs in smart farms. Then, two main approaches to configure solutions for UGVs in agricultural tasks are described: the automation of conventional vehicles and specifically designed mobile platforms. Their advantages and shortcomings regarding their working features are highlighted. This material enables the definition of other operating characteristics of UGVs to meet the smart farm requirements. Finally, the last section presents some conclusions.

2. UGV for agriculture

Ground mobile robots, equipped with advanced technologies for positioning and orientation, navigation, planning, and sensing, have already demonstrated their advantages in outdoor applications in industries such as mining [3], farming, and forestry [4, 5]. The commercial availability of GNSS has provided easy ways to configure autonomous vehicles or navigation systems to assist drivers in outdoor environments, especially in agriculture, where many highly accurate vehicle steering systems have become available [6, 7]. These systems aid operators in the precise guidance of tractors using LIDAR (light/laser detection and ranging) or GNSS technology but do not endow a vehicle or tool with any level of autonomy. Nevertheless, other critical technologies must also be incorporated to configure UGVs, such as the safety systems responsible for detecting obstacles in the robots' path and safeguarding humans and animals in the robots' surroundings as well as preventing collisions with obstacles or other robots. Finally, robot communications with operators and external servers (cloud technologies) through wireless communications that include the use of cyber-physical systems (CPSs) [8] and Internet of things (IoT) [9] techniques will be essential to incorporate decision-making systems based on big data analysis. Such integration will enable the expansion of decision processes into fields such as machine learning and artificial intelligence. Smart factories are based on the strongly intertwined concepts of CPS, IoT, big data, and cloud computing, and UGVs for smart farms should be based on the same principles to minimize the traditional delays in applying the same technologies to industry and agriculture.

The technology required to deploy more robotic systems into agriculture is available today, as are the clear economic and environmental benefits of doing so. For example, the global market for mobile robots, in which agricultural robots are a part, is expected to increase at a compound annual growth rate of over 15% from 2017 to 2025, according to recent forecast reports [10]. Nevertheless, manufacturers of agricultural machinery seem to be reluctant to commercialize fully robotic systems, although they have not missed the marketing potential of showing concepts [11, 12]. In any event, according to the Standing Committee on Agricultural Research [13], further efforts should be made by both researchers and private companies to invent new solutions.

Most of the robotics and automation systems that will be used in precision agriculture—including systems for fertilizing, planting, spraying, scouting, and harvesting (**Figure 1**)—will require the coordination of detection devices, agricultural implements, farm managing systems, and UGVs. Thus, several research groups and companies have been working on such systems. Specifically, two trends can be identified in the development of UGVs: the automation of conventional agricultural vehicles (tractors) and the development of specifically designed mobile platforms. The following sections discuss these two types of vehicles.

3. Automation of conventional vehicles

The tractor has been the central vehicle for executing most of the work required in a crop field. Equipped with the proper accessories, this machine can till, plant, fertilize, spray, haul, mow, and even harvest. Their adaptability to dissimilar tasks

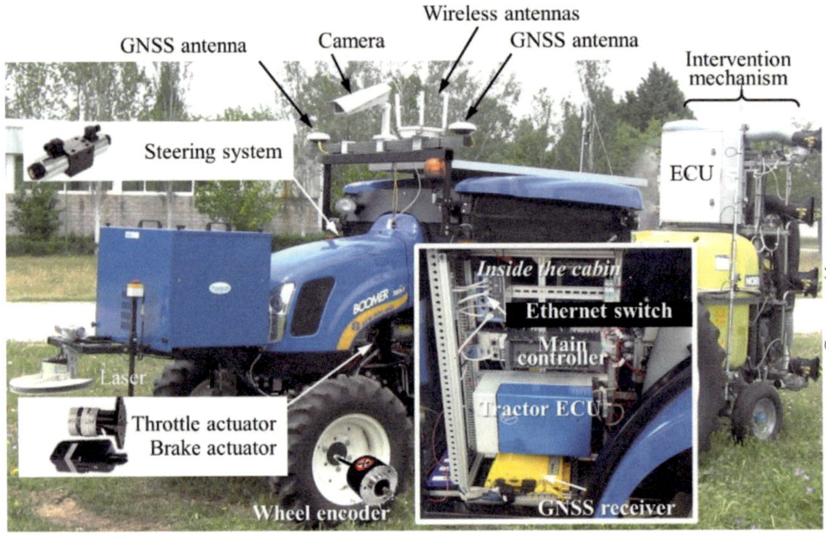

Figure 2.
An example of agricultural tractor automation-distribution of sensorial and actuation systems for transforming an agricultural tractor into a UGV (Gonzalez-de-Santos et al., 2017).

makes tractors a prime target for automation, which would enable productivity increases, improve safety, and reduce operational costs. **Figure 2** shows an example of the technologies and equipment for automating agricultural tractors.

Numerous worldwide approaches to automating diverse types of tractors have been researched and developed since 1995 when the first GNSS was made available to the international civilian community of users, which opened the door for GPS-guided agricultural vehicles (auto-steering) and controlled-traffic farming.

The first evaluations of GPS systems for vehicle guidance in agriculture were also published in 1995 [14] demonstrating its potential and encouraging many research groups around the world to automate diverse types of tractors. The earliest attempts were made at Stanford University in 1996, where an automatic control system for an agricultural tractor was developed and tested on a large farm [15]. The system used a location system with four GPS antennas. Around the same time, researchers at the University of Illinois, USA, developed a guidance system for an autonomous tractor based on sensor fusion that included machine vision, real-time kinematics GPS (RTK-GPS), and a geometric direction sensor (GDS). The fusion integration methodology was based on an extended Kalman filter (EKF) and a two-dimensional probability-density-function statistical method. This system achieved a lateral average error of approximately 0.084 m at approximately 2.3 m s^{-1} [16].

A few years later, researchers at Carnegie Mellon University, USA, developed some projects that made significant contributions. The Demeter project was conceived as a next-generation self-propelled hay harvester for agricultural operations, and it became the most representative example of such activity [17]. The positional data was fused from a differential GPS, a wheel encoder (dead reckoning), and gyroscopic system sensors. The project resulted in a system that allowed an expert harvesting operator to harvest a field once, thus programming the field. Subsequently, an operator with lesser skill could "playback" the programmed field at a later date. The semi-autonomous agricultural spraying project, developed by the same research group, was devoted to making pesticide spraying significantly cheaper, safer, and more environmentally friendly [18]. This system enabled a remote operator to oversee the nighttime operation of up to four spraying vehicles. Another example is research conducted at the University of Florida, USA, [19], in which two individual autonomous guidance systems for use in a citrus grove were developed and tested along curved paths at a speed of approximately 3.1 m s^{-1}. One system, based on machine vision, achieved an average guidance error of approximately 0.028 m. The other system, based on LIDAR guidance, achieved an average error of approximately 0.025 m.

Similar activities started in Europe in the 2000s. One example is the work performed at LASMEA-CEMAGREF, France, in 2001, which evaluated the possibilities of achieving recording-path tracking using a carrier phase differential GPS (CP-DGPS), as the only sensor. The vehicle heading was derived according to a Kalman state reconstructor and a nonlinear velocity independent control law was designed that relied on chained systems properties [20].

A relevant example of integrating UGVs with automated tools is the work conducted at the University of Aarhus and the University of Copenhagen, Denmark [21]. The system comprised an autonomous ground vehicle and a side shifting arrangement affixed to a weeding implement. Both the vehicle and the implement were equipped with RTK-GPS; thus, the two subsystems provided their own positions, allowing the vehicle to follow predefined GPS paths and enabling the implement to act on each individual plant, whose positions were automatically obtained during seeding.

Lately, some similar automations of agricultural tractors have been conducted using more modern equipment [22, 23], and some tractor manufacturers have already presented noncommercial autonomous tractors [11, 12]. This tendency to automate existing tractors has been applied to other types of lightweight vehicles for specific tasks in orchards such as tree pruning and training, blossom and fruit thinning, fruit harvesting, mowing, spraying, and sensing [24]. **Table 1** summarizes the UGVs based on commercial vehicles for agricultural tasks.

Institution	Year	Description
Stanford University (USA) [15]	1996	Automatic large-farm tractor using 4 GPS antennas
University of Illinois (USA) [16]	1998	A guidance system using a sensor based on machine vision, an RTK-GPS, and a GDS
Carnegie Mellon University (USA)—Demeter project [17]	1999	A self-propelled hay harvester for agricultural operations
Carnegie Mellon University (USA)—Autonomous Agricultural Spraying project [18]	2002	A ground-based vehicles for pesticide spraying
LASMEA-CEMAGREF (France) [20]	2001	This study investigated the possibility of achieving vehicle guiding using a CP-DGPS as the only sensor
University of Florida (USA) [19]	2006	An autonomous guidance system for citrus groves based on machine vision and LADAR
University of Aarhus and the University of Copenhagen (Denmark) [21]	2008	An automatic intra-row weed control system connected to an unmanned tractor
RHEA consortium (EU) [22]	2014	A fleet (3 units) of tractors that cooperated and collaborated in physical/chemical weed control and pesticide applications for trees
Carnegie Mellon University (USA) [24]	2015	Self-driving orchard vehicles for orchard tasks
University of Leuven (Belgium) [23]	2015	Tractor guidance using model predictive control for yaw dynamics

Table 1.
UGVs based on commercial vehicles.

Nevertheless, UGVs suitable for agriculture remain far from commercialization, although many intermediate results have been incorporated into agricultural equipment—from harvesting to precise herbicide application. Essentially, these systems are installed on tractors owned by farmers and generally consist of a computer (the controller), a device for steering control, a localization system (mostly based on RTK-GPS), and a safety system (mostly based on LIDAR). Many of these systems are compatible only with advanced tractors that feature ISOBUS control technology [25], through which controllers connected to the ISOBUS can access other subsystems of the tractor (throttle, brakes, auxiliary valves, power takeoff, linkage, lights, etc.). Examples of these commercial systems are AutoDrive [26] and X-PERT [27].

An important shortcoming of these solutions is their lack of intelligence in solving problems, especially when obstacles are detected because they are not equipped with technology suitable for characterizing and identifying the obstacle type. This information is essential when defining any behavior other than simply stopping and waiting for the situation to be resolved. Another limitation of this approach is that the conventional configuration of a standard tractor driven by an operator is designed to maximize the productivity per hour; thus, the general architecture of the system (tractor plus equipment) is only roughly optimized.

4. Specifically designed mobile platforms

The second approach to the configuration of mobile robots for agriculture is the development of autonomous ground vehicles with specific morphologies, where researchers develop ground mobile platforms inspired more by robotic principles than by tractor technologies. These platforms can be classified based on their locomotion system. Ground robots can be based on wheels, tracks, or legs. Although legged robots have high ground adaptability (that enables the vehicles to work on irregular and sloped terrain) and intrinsic omnidirectionality (which minimizes the headlands and, thus, maximizes croplands) and offer soil protection (discrete points in contact with the ground that minimize ground damage and ground compaction, an important issue in agriculture), they are uncommon in agriculture; however, legged robots provide extraordinary features when combined with wheels that can configure a disruptive locomotion system for smart farms. Such a structure (which consists of legs with wheels as feet) is known as a wheel-legged robot. The following sections present the characteristics, advantages, and disadvantages of these specifically designed types of robots.

4.1 Wheeled mobile robots

4.1.1 Structures of wheeled robots

The structure of a wheeled mobile platform depends on the following features:

Number of wheels: Three nonaligned wheels are the minimum to ensure platform static stability. However, most field robots are based on four wheels, an approach that increases the static and dynamic stability margins [28].

Wheel orientation type: An ordinary wheel can be installed on a platform in different ways that strongly determine the platform characteristics. Several wheel types can be considered:

a. **Fixed wheel**: This wheel is connected to the platform in such a way that the plane of the wheel is perpendicular to the platform and its angle (orientation) cannot change.

b. **Orienting wheel**: The wheel plane can change its orientation angle using an orientation actuator.

c. **Castor wheel**: The wheel can rotate freely around an offset steering joint. Thus, its orientation can change freely.

Wheel power type: Depending on whether wheels are powered, they can also be classified as follows:

a. Passive wheel: The wheel rotates freely around its shaft and does not provide power.

b. Active wheel: An actuator rotates the wheel to provide power.

Wheel arrangement: Different combinations of wheel types produce mobile platforms with substantially different steering schemes and characteristics.

a. Coordinated steering scheme: Two fixed active wheels at the rear of the platform coupled with two passive orienting wheels at the front of the platform are the most common wheel arrangement for vehicles. To maintain all wheels in a pure rolling condition during a turn, the wheels need to follow curved paths with different radii originating from a common center [29]. A special steering mechanism, the Ackermann steering system, which consists of a 4-bar trapezoidal mechanism (**Figure 3a**), can mechanically manage the angles of the two steering wheels. This system is used in all the vehicles presented in **Table 2**. It features medium mechanical complexity and medium control complexity. One advantage of this system is that a single actuator can steer both wheels. However, independent steering requires at least three actuators for steering and power (**Figure 3b**).

b. Skid steering scheme: Perhaps the simplest structure for a mobile robot consists of four fixed, active wheels, one on each corner of the mobile platform. Skid steering is accomplished by producing a differential thrust between the left

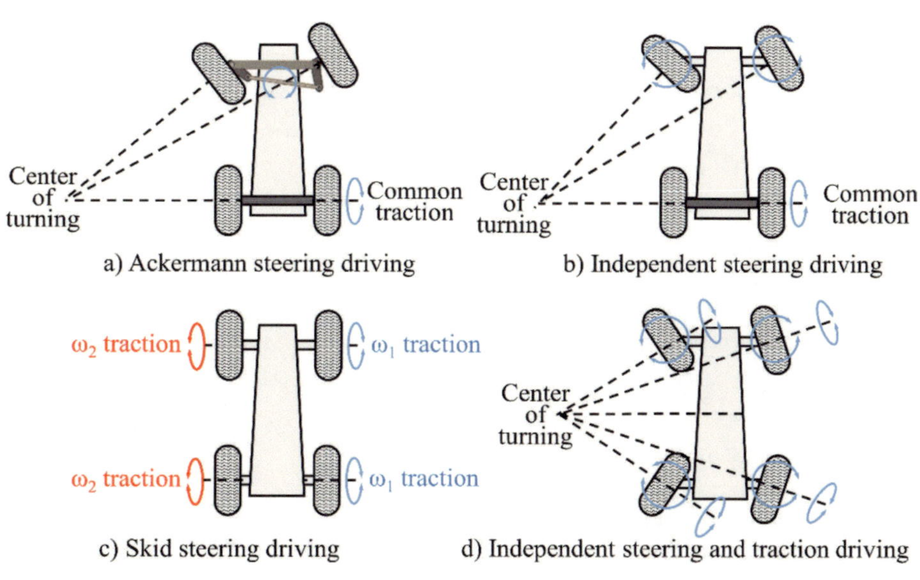

Figure 3.
Steering driving systems: (a) Ackermann steering system; (b) independent steering; (c) skid steering system and (d) independent steering and traction system.

and right sides of the vehicle, causing a heading change (**Figure 3c**). The two wheels on one side can be powered independently or by a single actuator. Thus, the motion of the wheels in the same direction produces backward/forward platform motion; and the motion of the wheels on one side in the opposite direction to the motion of wheels on the other side produces platform rotation.

c. Independent steering scheme: An independent steering scheme controls each wheel, moving it to the desired orientation angle and rotation speed (**Figure 3d**). This steering scheme makes wheel coordination and wheel

Steering scheme	**Characteristics**
Coordinated	Advantages: • Simplicity. • Few actuators (2) if based on the Ackermann device. • Good turning accuracy if the front wheels are steered independently. Disadvantages: • Large turning radii. • Ideal rotation in only three steering angles if based on the Ackermann device. • Requires three actuators and more complex control algorithms if based on front wheels steered independently. • Steering control on loose grounds, e.g., after plowing, is difficult. Use in smart farms: • New mobile robotic designs are abandoning this scheme, which only offers simplicity. Hence, such steering control is not expected to be used in smart farms.
Skid	Advantages: • Compact size, robustness, few parts. • Agility (motion with heading control and zero-radius turns). • Few actuators (2). Disadvantages: • The maximum forward thrust is not maintained during turns. • Terrain irregularities and tire-soil effects demand unpredictable power supply. • Vehicle rotations erode the ground and wore the tires. Use in smart farms: • This steering scheme is simple and robust, but not very precise in loose terrain; hence, it could be used in smart farms, e.g., for indoor tasks, but not for infield tasks.
Independent	Advantages: • Full mobility (including crab motion). Disadvantages: • Many actuators and parts (eight for a four-wheel robot). • Complex control algorithms. Use in smart farms: • This steering scheme is the more versatile of the schemes, but it is also more complex and expensive. However, most of the engineering systems evolve by increasing their sophistication and robustness while decreasing their cost; hence, this scheme will be intensively used in smart farms.

Table 2.
Characteristics of wheeled structures.

position accuracy more complex but provides some advantages in maneuverability. In addition, this scheme provides crab steering (sideways motion at any angle α; $0 \leq \alpha \leq 2\pi$) by aligning all wheels at an angle α with respect to the longitudinal axis of the mobile platform. Finally, the coordination of driving and steering results in more efficient maneuverability and reduces internal power losses caused by actuator fighting. The independent steering scheme requires eight actuators for a four-wheel vehicle.

Table 2 summarizes the advantages and drawbacks of these schemes. Note that the number of actuators increases the total mass of a robot as well as its mechanical and control complexity (more motors, more drivers, more elaborate coordinating algorithms, etc.).

4.1.2 Examples of wheeled robots

Some examples of wheeled mobile platforms for agriculture are the conventional tractor using the Ackermann steering system (**Figure 2**) with two front passive and steerable wheels and two rear fixed and active wheels.

Skid steering platforms can be found in many versions. For example,

- Four fixed wheels placed in pairs on both sides of the robot

- Two fixed tracks, each one placed longitudinally at each side of the robot,

- Two fixed wheels placed at the front of the robot and two castor wheels placed at the rear (**Figure 4c**), etc.

Regarding the independent steering scheme, the robot developed by Bak and Jakobsen [30] is one of the first representative examples (**Figure 4a**). This platform was designed specifically for agricultural tasks in wide-row crops and featured good ground clearance (approximately 0.5 m) and 1-m wheel separation. The platform is based on four-identical wheel modules. Each one includes a brushless electric motor that provides direct-drive power, and steering is achieved by a separate motor.

An example of a mobile platform under development that focuses on performing precision agricultural tasks is AgBot II (**Figure 4c**). This is a platform that follows the skid steering scheme with two front fixed wheels (working in skid or differential mode) and two rear caster wheels. It is intended to work autonomously on both large-scale and horticultural crops, applying fertilizer, detecting and classifying weeds, and killing weeds either mechanically or chemically [31, 32]. Another robot is Robot for Intelligent Perception and Precision Application (RIPPA), which is a light, rugged, and easy-to-operate prototype for the vegetable growing industry. It is used for autonomous high-speed, spot spraying of weeds using a directed micro-dose of liquid when equipped with a variable injection intelligent precision applicator [33]. Another example is Ladybird (**Figure 4b**), an omnidirectional robot powered with batteries and solar panels that follows the independent steering scheme. The robot includes many sensors (i.e., hyperspectral cameras, thermal and infrared detecting systems, panoramic and stereovision cameras, LIDAR, and GPS) that enable assessing crop properties [34]. One more prototype, very close to commercialization, is Kongskilde Vibro Crop Robotti, which is a self-contained track-based platform that uses the skid steering scheme. It can be equipped with implements for precision seeding and mechanical row crop cleaning units. This robot can work for 2–4 hours at a 2–5 km h^{-1} rate and is supplied by captured electric energy [35].

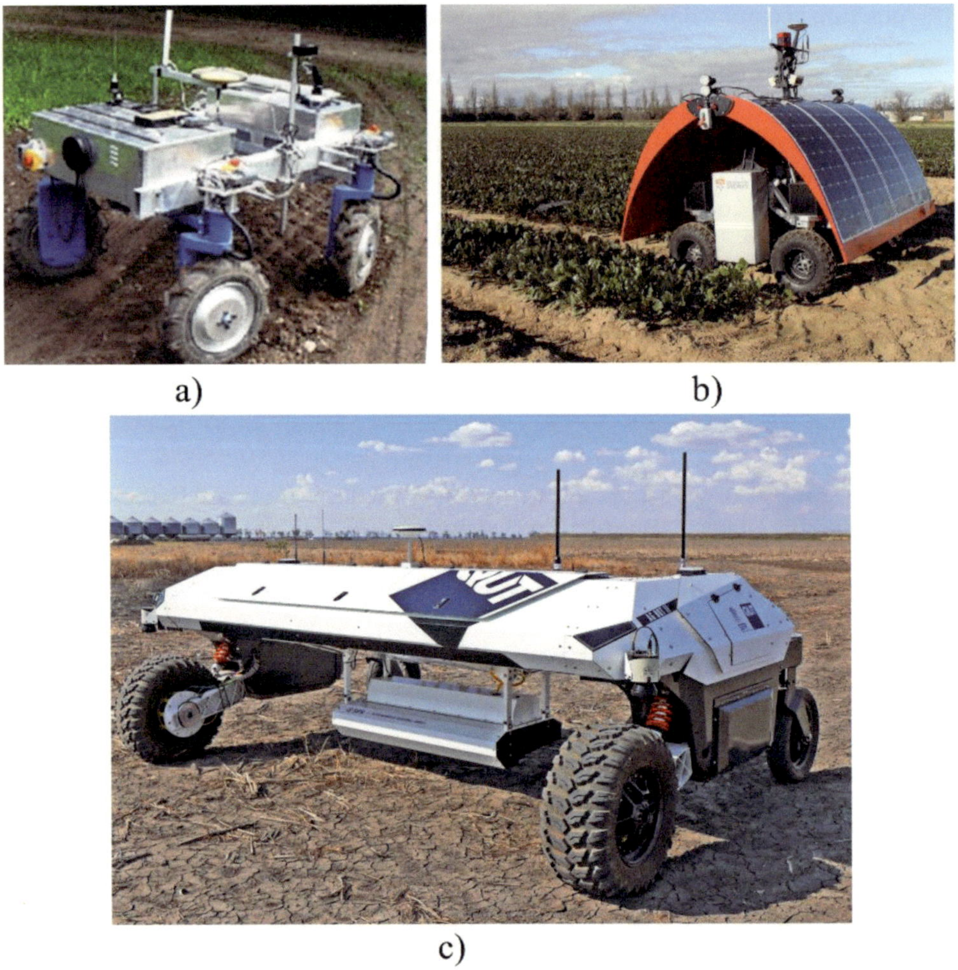

Figure 4.
Pictures of several specifically-designed agricultural platforms. (a) Robot for weed detection, courtesy of T. Bak, Department of Agricultural Engineering, Danish Institute of Agricultural Sciences; (b) ladybird, courtesy of J. P. Underwood, Australian Centre for Field Robotics at the University of Sydney [34]; (c) AgBot II, courtesy of O. Bawden, strategic Investment in Farm Robotics, Queensland University of Technology [31].

These robots are targeted toward fertilizing, seeding, weed control, and gathering information, and they have similar characteristics in terms of weight, load capacity, operational speed, and morphology. Tools, instrumentation equipment, and agricultural implements are connected under the robot, and tasks are performed in the area just below the robot, which optimizes implement weight distribution. These robots have limitations for use on farmland with substantial (medium to high) slopes or gully erosion. Nevertheless, some mobile platforms are already commercially available. Two examples of these vehicles are the fruit robots Cäsar [36] and Greenbot [37].

Cäsar is a remote-controlled special-purpose vehicle that can perform temporarily autonomous operations in orchards and vineyards such as pest management, soil management, fertilization, harvesting, and transport. Similarly, Greenbot is a self-driving machine specially developed for professionals in the agricultural and horticultural sectors who perform regular, repetitious tasks. This vehicle can be used not only for fruit farming, horticulture, and arable farming but also in the urban sector and even at waterfronts or on roadsides.

Despite their current features, the existing robots lack flexibility and terrain adaptability to cope with diverse scenarios, and their safety features are limited. For example:

- They focus only on orchard and vineyard activities.
- They have ground clearance limitations.
- They are unsuitable for rough terrain or slopes.
- They must be manually guided to the working area rather than freely and autonomously moving to different working areas around the farm.
- They possess no advanced detection systems for weed or soil identification, which limits their use to previously planned tasks related to selective treatment.
- They lack dynamic safety systems capable of recognizing or interpreting safety issues; thus, they are incapable of rescheduling or solving problems by themselves.

In addition, existing UGVs for agriculture lack communication mechanisms for providing services through cloud technologies, CPS, and IoT techniques, crucial instruments to integrate decision-making systems based on big data analysis, as is being done in the smart factory concept.

Table 3 summarizes the diverse robotic platforms, and **Figure 4** depicts some of these platforms.

4.2 Wheel-legged robots

4.2.1 Structures of wheel-legged robots

The structure of a wheel-legged mobile platform depends on (i) the number of legs, (ii) the leg type, and (iii) the leg arrangement. The feet consist of 2-DOF steerable powered wheels as illustrated in **Figure 5**.

Number of legs: The minimum number of legs required for statically stable walking is four-three legs providing support in the form of a stable tripod while the other leg performs the transference phase [38]. Combining sequences of leg

Vehicle	Type*	Year	Description
AgBot II [32]	P	2014	A platform that follows the skid steering scheme with two front fixed wheels (working in skid or differential mode) and two rear caster wheels
Ladybird [34]	P	2015	An omnidirectional robot powered with batteries and solar panels that uses the independent steering scheme
Greenbot [37]	C	2015	A self-driving robot for tasks in agriculture and horticulture
Cäsar [36]	P	2016	A remotely controlled platform for temporary, autonomous use in fruit plantations and vineyards
RIPPA [33]	P	2016	A light, rugged, and easy-to-operate prototype for the vegetable growing industry
Vibro Crop Robotti [35]	C	2017	A self-contained track-based platform that uses the skid steering scheme

* *P-prototype; C-commercial.*

Table 3.
Robots designed specifically for agriculture.

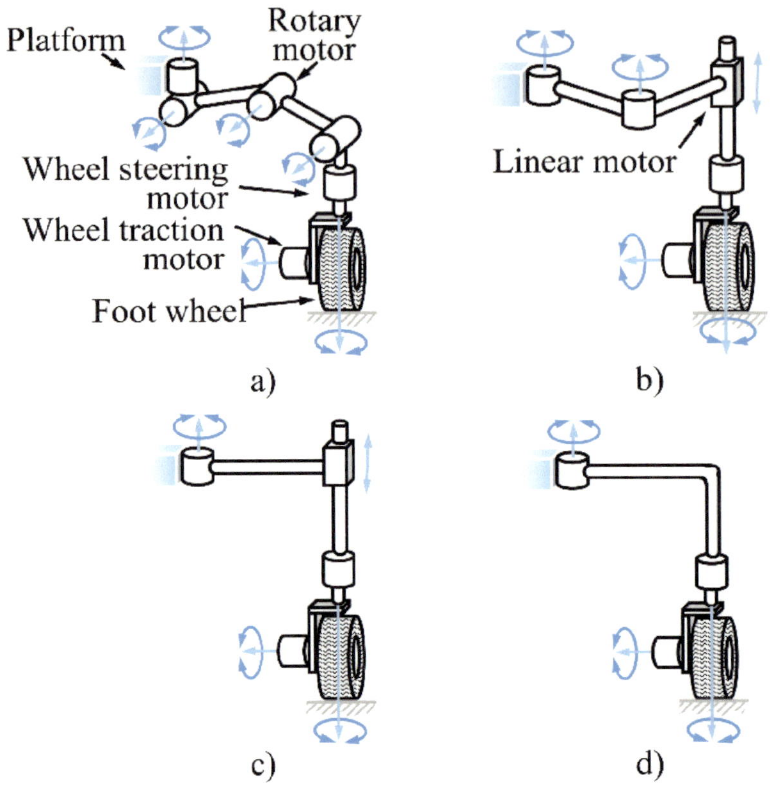

Figure 5.
Wheel-legged structures. (a) 4-DOF articulated leg; (b) 3-DOF SCARA leg; (c) 2-DOF SCARA leg; (d) 1-DOF leg.

transferences with stable tripods produce a walking motion. A wheel-legged robot requires only three legs for translational motion, which provides additional terrain adaptation.

Leg type: Legs are based on the typical configurations of manipulators; thus, articulated, cylindrical, Cartesian, and pantographic configurations are the types used most often.

Leg arrangement: The normal arrangement for a $2n$-legged robot is to distribute n legs uniformly on the longitudinal sides. Four-legged structures present some advantages regarding terrain adaptability, ground clearance, and track width control (crop adaptability) but also have some drawbacks, such as additional mechanical complexity (complex joints designs, including actuators and brakes) and control of redundant actuated systems, which exhibit complex interactions with the environment and make motion control more difficult than that of conventional wheeled platforms. **Table 4** illustrates different theoretical wheel-legged structures.

4.2.2 Examples of wheel-legged robots

Figure 6a illustrates the structure scheme of a wheel-legged robot based on the 3-DOF SCARA leg (See **Figure 5b**) with full terrain adaptability, ground clearance control, crop adaptability, and capability of walking, and **Figure 6b** shows the structure of a wheel-legged robot exhibiting full terrain adaptability, ground clearance control, and crop adaptability; however, it cannot walk under static stability.

Another interesting example is the structure of BoniRob [39], a real wheel-legged platform for multipurpose agriculture applications, which consists of four independently steerable powered wheeled legs with the structure illustrated in **Figure 5d** (1-DOF legs with a 2-DOF wheeled foot). This robot can adjust the distance between its wheel sets, making it adaptable to many agricultural scenarios. The platform can be equipped with common sensorial systems used in robotic agricultural applications, such as LIDAR, inertial sensors, wheel odometry, and GPS. Moreover,

Structure	Characteristics
A 4-DOF articulated leg with a 2-DOF wheeled foot (**Figure 5a**)	Advantages: • Full terrain adaptability and ground clearance control. • Crop control. • Full capability for walking. Disadvantages: • A huge number of actuators (24) that jeopardize the robot's reliability. Use in smart farms: • This structure is the most complex structure that exhibits complete wheel positioning and orientation in its working volume. However, the orientation of the wheel does not provide additional characteristics regarding stability or traction. Thus, this structure provides the same advantages as other structures (see **Figure 5c**) but with extra complexity, which will jeopardize its application in smart farms. This structure is presented here as the most complex platform.
A 3-DOF motion-decoupled leg* with a 2-DOF wheeled foot (**Figure 5b**)	Advantages: • Full terrain adaptability and ground clearance control. • Crop adaptability. • Full capability for walking. Disadvantages: • A large number of actuators (20). Use in smart farms: • This structure provides full positioning of the wheel in its working volume and can control the robot's body leveling, which allows for the wheel plane to be aligned with gravity, which provides an excellent robot's stability using fewer motors than the structure illustrated in **Figure 5a**. In addition, this structure can walk under static stability, an interesting feature when the robot works in very irregular, soft, or muddy terrain. Its terrain adaptability, ground clearance control, and crop adaptability, along with its medium complexity, make this structure the most promising for use in smart farms in the long term.
A 2-DOF motion-decoupled leg* with a 2-DOF wheeled foot (**Figure 5c**)	Advantages: • A medium number of actuators (16). • Full terrain adaptability and ground clearance control. • Crop adaptability. Disadvantages: • Limitations for walking. Use in smart farms: • This structure can control the ground clearance, leveling, and distance between wheels; the latter determines the adaptation to different crops (distance between crop rows). Nevertheless, the wheel moves on a vertical-cylindrical surface rather than in a working volume. This fact impedes the robot from walking and, thus, exhibits worse characteristics than the structure illustrated in **Figure 5b**. In any case, it can be a proper structure to introduce wheel-legged vehicles and could be used in the short term.

Structure	Characteristics
A 1-DOF leg with a 2-DOF wheeled foot (**Figure 5d**)	Advantages: • A small number of actuators (12). • Crop adaptability. Disadvantages: • No terrain adaptability or ground clearance control. • Limitations for walking. Use in smart farms: • This structure has no capabilities for walking or controlling the ground clearance of the vehicle or its leveling. However, the structure is simple and could be used as an introductory robot structure for smart farms in the short term.

**Cylindrical, Selective Compliant Articulated Robot Arm (SCARA) or Cartesian.*

Table 4.
Wheel-legged structures.

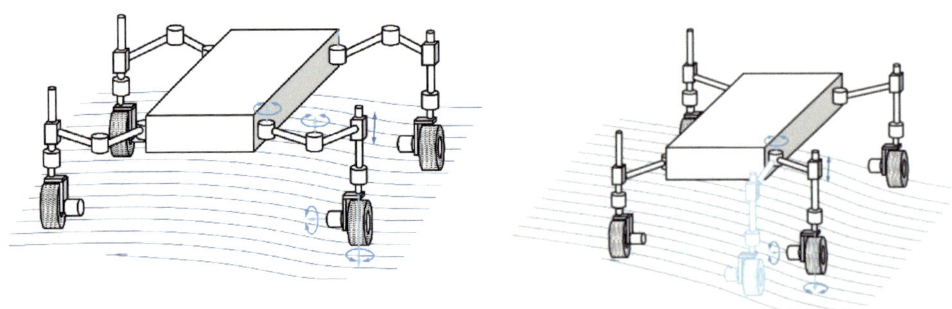

a) b)

Figure 6.
Model of wheel-legs: (a) full terrain-crop adaptability, (b) full terrain and partial crop adaptability.

the robotic platform can be retrofitted and upgraded with swappable application modules or tools for crop and weed identification, plant breeding applications, and weed control. This robotic platform is completely powered by electricity, which is more environmentally friendly but reduces its operational working time compared to conventional combustion-engine systems. Nevertheless, this robot configuration requires custom-built implements, which prevent the reuse of existing implements and, thus, jeopardize the introduction of this robot to the agricultural market.

5. Smart farm UGV characteristics

In addition to their needed characteristics for infield operations, the robots fulfilling the demands of a smart farm will require the operating requirements summarized in the following paragraphs and **Table 5**.

Small size: The idea that using small robots provides many advantages over the use of conventional large vehicles has been widely discussed over the past decade [22, 40]. It is broadly accepted that although several small robots can cost the same as a large machine and accomplish the same amount of work, using small robots allows a multi-robot system to continue a task even if a number of robots fail (re-planning the task). Moreover, the reduced weight of the small robots reduces terrain compaction and allows farmers to acquire robots incrementally.

Characteristics	Value
Dimensions	Length: ~3.0 m; width: ~1.50 m; height: ~1.00 m
Weight	1200–1700 kg
Payload	500–1000 kg
Comments: These characteristics are estimations based on the current medium-sized vehicles reported in this chapter that are capable of carrying agricultural implements. Robots for carrying sensing systems can be truly small (low payloads), but vehicles for treatments need to carry medium to heavy loads (pesticides, fertilizes, etc.). For example, existing sprayers [45] weigh approximately 600–700 kg including 200–300 L of active ingredient.	
Speed	3–25 km h^{-1}
Comments: Treatment speed is limited by the treatment process that depends on physical laws. However, robots need to move among working fields minimizing moving time; therefore, they must feature a reasonably high top speed.	
Position accuracy	±0.02 m
Comments: The current DGPS accuracy seems to be sufficient for real applications. However, specific real-time localization systems, RTLS, can be used in small areas where GNSS is unavailable (radio frequency identification tags (RFID), ultra-wide band tags (UWB), etc.). These technologies will be essential in smart farms to ensure positioning precision in GNSS occluded areas.	
Clearance	0.35–1 m
Comments: Weed control is performed at an early crop-growth stage; therefore, the minimum ground clearance of the robot must be approximately 0.35 m. A ground clearance of approximately 1 m will facilitate application of treatments at later crop-growth stages. The ideal approach would be to control the ground clearance to optimize the working height of the implements based on the crop. Existing robots cannot control their ground clearance, but some wheel-legged configurations can meet this specification (**Figure 5a, b**, and **c**).	
Track width	1.50–2.25 m
Comments: To preserve crops in narrow-row situations, a tramline control is required; however, in wide-row crops, the tramlines must be located in the inter-row spacing. Taking maize as an example, which is planted at an inter-row spacing of approximately 0.75 m in some areas in Europe, a robot track width of 1.50 to 2.25 m is required to enable 2 or 3 rows to pass under the robot's body. Controlling robot track width is imperative in a smart farm world. This characteristic is exhibited by wheeled-legged robots, which makes them a good candidate for UGVs in smart farms.	
Energetic autonomy	~10 h
Comments: Robots based on combustion engines (e.g., tractors) can operate autonomously for approximately 10 hours, at minimum. The duration of autonomous operation for electrically driven systems should be similar. Some existing prototypes already meet this expectation [31]. In any case, the increasing improvement in battery technology will enlarge the energetic autonomy of future vehicles and robots.	

Table 5.
Prospective characteristics for UGVs in smart farms.

Flexibility: Agricultural robots must be capable of adapting to many different scenarios (e.g., crops, row types, etc.) and tasks (e.g., plow, sow, fumigate, etc.). Thus, the robots must also be able to accommodate different agricultural implements, which should attach to or connect to (respectively, detach or disconnect from) the robots automatically.

Although conventional tractors are proven and highly reliable machines, they lack some adaptability features. Tractors have normally fixed distances between wheels, which makes them unsuitable for working on crops with different distances between rows. Using mobile platforms capable of controlling the distance between wheels could alleviate this problem, allowing the machines to adapt to different crops under different situations.

Maneuverability: Robots must be capable of performing small radius turns while adapting to different terrain. This last feature requires independent vertical control of wheels with respect to the robot's body.

A steering system capable of zero-radius turns would be a proper solution, and this feature can be implemented by different structures as discussed in the previous section. Thus, minimization of headlands and wheel distance control can be achieved using either conventional or new articulated structures. Among the conventional structures, the skid steering scheme based on wheels or tracks is capable of zero-radius turns without additional steering mechanism, which helps in minimizing the headlands. However, separating and controlling the distance between contralateral wheels/tracks requires an active system (which already exists for some tracked vehicles used in the building industry).

Mobile platform structures based on coordinated or independent steering schemes can achieve zero-radius turns, but they still lack intrinsic track width control and require additional mechanisms. Another structure is the wheel-legged mechanism. Legged robots exhibit high terrain adaptability on irregular ground, but wheeled robots have speed advantages on smooth terrain; that is, they complement each other. Therefore, the most complete wheel-legged mechanism (**Figure 6a**) is a leg with three degrees of freedom [38] with an active wheel as a foot, where the wheel is steered and driven separately. This is a disruptive design not verified yet that will provide extraordinary characteristics to robots for smart farm applications. Thus, the wheels drive and steer, while the legs provide track-width control and terrain adaptation, i.e., they control the robot's body leveling and ground clearance. This is the most capable system regarding ground clearance and body pose control, but it comes at the cost of higher mechanical complexity. Nevertheless, intermediate solutions can be developed to reduce the number of actuators while maintaining appropriate robot characteristics. **Table 4** summarizes different wheel-legged theoretical solutions indicating advantages and shortcomings, and **Figure 5** shows some sketches of practical solutions.

Resilience: Resilience is the ability to recover from malfunctions or errors. Initializing complex robots is a time-consuming procedure, especially when several robots are collaborating on the same task. Agricultural mobile robots must be resilient enough to ensure profitability. Thus, they must be easily shut down and started up (essential for error recovery); moreover, they must facilitate changing between manual operation mode and autonomous operation mode and vice versa.

Efficiency: UGV should be more efficient than conventional, manned solutions. This can be accomplished by systems that:

- Minimize energy consumption by optimizing the robot trajectories during the mission

- Drastically reduce the use of herbicides and fertilizers by using intelligent detection systems, tools, and decision-making algorithms

- Eliminate the need for a driver and minimize operator risk

- Minimize unnecessary crop damage and soil compaction

Friendly human-machine interfaces (HMI): A friendly interface is required to facilitate the introduction of robots into agriculture and to achieve profitability. Intuitive, reliable, comfortable, and safe HMIs are essential for farmers to accept robotic systems. The HMIs should be implementable on devices such as smartphones and tablets.

Communications: Communications in the smart farm must capitalize on CPS and IoT to collect sufficient data to take advantage of the big data techniques and enable communication with the cloud for use via different services (software as a service, platform as a service, and infrastructure as a service) offered by cloud providers [41].

Wireless communications with the operator and/or a central controller for control commands and data exchanges, including images and real-time video, will be required. Wireless communication among robots will also be required for coordination and collaboration.

Standardization of mechanical and electrical/electronic interfaces: Commercial equipment must comply with well-defined standards and homologous procedures before adoption by industry. Subsystems such as LIDAR units, computers, and wireless or Internet communication (4G/5G) devices and GNNS receivers and antennas are already off-the-shelf components, but mobile platforms must also cope with some standards related to agricultural machinery [25, 42].

Safety: Safety systems for agricultural robots must focus on three stages: (i) safety to humans, (ii) safety to crops, and (iii) safety to the robots themselves.

Safety for humans and robots can usually be accomplished through a combination of computer vision, LIDAR, and proximity sensors to infer dangerous situations and halt robot motion, whereas safety to crops is achieved through precise steering that guides the robot to follow the crop rows accurately using the crop position acquired at seeding time or real-time crop-detection systems. Following these three stages, a step forward in safety for agricultural robots would be the integration of a two-level safety system relying on the following:

- **A low-level safety system** that detects short-range obstacles with the purpose of avoiding imminent collisions. This level should be implemented within the robot controller and based on commercial components.

- **A high-level safety system** that detects and discriminates obstacles at an adequate distance to allow the robotic system to make decisions (i.e., re-planning a trajectory). This level should include vision, infrared, and hyperspectral cameras that provide information about the surroundings. Optical flow methods should be applied to detect obstacles in motion and compute their speed and direction to predict potential collisions [43]. Hence, optical sensors should track obstacles and their movements, dynamically compute safe zones, and adjust a robot's speed and direction of movement according to the given situation.

Regardless of the exact approach, standards on safety machinery must be taken into consideration [42] to ensure that systems will meet regulations and will be able to achieve certification.

Environmentally friendly impact: Both intervention mechanisms (implements) and mobile robots must be environmentally friendly (e.g., use fewer chemicals and cause less soil compaction) while improving the efficiency of the agricultural processes (i.e., reduce chemical costs while equaling or improving production). In addition, current agricultural vehicles use fossil fuels that emit large amounts of pollutants into the air such as carbon dioxide (CO_2), nitrogen oxide (NOX), carbon monoxide (CO), and hydrocarbon (HC) [44]. Furthermore, fuel can be spilled onto the ground, which is a long-term pollutant. These elements alter the environment and damage the ecosystem. One possible solution—envisaged as the likely future solution—is the use of electric vehicles.

Implements: The use of the conventional three-point hitch to attach implements to tractors should be changed as robots are introduced into agriculture. Instead, implements should be aligned with the robot's center of gravity to optimize the payload distribution and minimize compaction. Mechanical attachment and electrical connection to the implement should be automated. The definition of these types of interfaces is a pending issue; nevertheless, an intermediate solution allowing the use of both new and conventional attachment devices (three-point hitch) will facilitate the gradual introduction of robotic systems into the agricultural sector. Obviously, developing new robots and adapting existing implements to a new attachment/connection system is the only way to introduce the robots to real applications.

HMI: An HMI for operators to communicate with robots should be implementable on portable equipment (smartphones, tablets, etc.). Operators will use such devices to send commands and receive responses and data. Moreover, an additional device—an emergency button that works using radio signals—must be provided to stop the robots from malfunctioning or unsafe situations. These interfaces must be true user-friendly devices to be operated by farmers rather than by engineers, which is a vital aspect for the introduction of robotics into agriculture, as it is for industry and services.

Autonomy: Two basic types of autonomies will be needed in smart farms: behavioral autonomy and operational autonomy. Behavioral autonomy is primarily associated with autonomous robots and relies on artificial intelligence techniques. It refers to the robot's ability to deal with uncertainty in its environment to accomplish a mission. Operational autonomy is associated with the tasks the robot has to accomplish autonomously to become a UGV, i.e., the tasks required for the robot to work continuously without human intervention: refueling or recharging (energetic autonomy, see **Table 5**), herbicide/pesticide refilling, implement attaching, and crop offloading. These tasks, which can be solved using current automatic techniques, are currently being done with human intervention and should be fully automated in the smart farms.

Based on the existing agricultural vehicles and robot prototypes, robots to be deployed in smart farms should meet also the characteristics presented in **Table 5**.

6. Conclusions

The world population is increasing rapidly, causing a demand for more efficient production processes that must be both safe and respect the ecosystem. Industry has already planned to meet production challenges in the coming decades by defining the concept of the smart factory; the agriculture sector should follow a similar path to design the concept of the smart farm: a system capable of optimizing its performance across a wide network, learning from new conditions in real time and adapting the system to them and executing the complete production process in an autonomous manner. Smart factory and smart farm concepts have many commonalities and include some common solutions, but some specific aspects of smart farms should be studied separately. For example, the design of UGVs for outdoor tasks in agriculture (field robots) presents specific characteristics worthy of explicit efforts.

This chapter focused on reviewing the past and present developments of UGVs for agriculture and anticipated some characteristics that these robots should feature for fulfilling the requirements of smart farms. To this end, this chapter presented and criticized two trends in building UGVs for smart farms based on (i) commercial vehicles and (ii) mobile platforms designed on purpose. The former has been useful for evaluating the advantages of UGV in agriculture, but the latter offers additional benefits such as increased maneuverability, better adaptability to crops, and improved adaptability to the terrain. Clearly, independent-steering and skid-steering systems provide the best maneuverability, but depending on their complexity, wheel-legged structures can provide similar maneuverability and improved adaptability to crops and terrain as well as increased stability on sloped terrain. For example, the 4-DOF articulated wheeled leg (**Figure 5a**) and the 3-DOF SCARA leg (**Figure 5b** and **6a**) exhibit the best features at the cost of being the most complex. Note that although both structures have the same maneuverability features and adaptability to crops and terrain (ground clearance, body leveling, etc.), the 3-DOF SCARA leg involves one fewer motor per leg, which decreases the price and weight and improves the reliability of the robot. However, the 2-DOF SCARA leg also exhibits useful features regarding maneuverability, adaptability to crops, and adaptability to terrain (ground clearance control and body leveling) while using fewer actuators (**Figure 5c** and **6b**). For agricultural tasks carried out on flat terrain, the 1-DOF leg with a 2-DOF wheeled foot provides sufficient maneuverability and adaptability to crops with very few actuators (leg structure as in **Figure 5d**).

However, these robots also require some additional features to meet the needs of the smart farm concept, such as the following:

i. Flexibility to work on very dissimilar scenarios and tasks.

ii. Maneuverability to perform zero-radius turns, crab motion, etc.

iii. Resilience to recover itself from malfunctions.

iv. Efficiency in the minimization of pesticide and energy usage.

v. Intuitive, reliable, comfortable, and safe HMIs attractive to nonrobotic experts to ease the introduction of robotic systems in agriculture.

vi. Wireless communications to communicate commands and data among the robots, the operator, and external servers for enabling CPSs, IoT, and cloud computing techniques to support services through the Internet.

vii. Safety systems to ensure safe operations to humans, crops, and robots.

viii. Environmental impact by reducing chemicals in the ground and pollutants into the air.

ix. Standards: operational robots have to meet the requirements and specifications of the standards in force for agricultural vehicles.

x. Implement usage: although specific onboard implements for UGV are appearing, the capability of also using conventional implements will help in the acceptation of new technologies by farmers and, hence, the introduction of new-generation robotic systems.

xi. Autonomy: both behavioral autonomy and operation autonomy. Regarding power supplies, automobiles worldwide will likely be electric vehicles powered by batteries within the next few decades; thus, agricultural vehicles should embrace the same solution.

Regardless of these characteristics, UGVs for smart farms have to fulfill the requirements of multi-robot systems, which is a fast-growing trend [22, 40, 46].

Multi-robot systems based on small-/medium-sized robots can accomplish the same work as a large machine, but with better positioning accuracy, greater fault tolerance, and lighter weights, thus reducing soil compaction and improving safety. Moreover, they can support mission coordination and reconfiguration. These capabilities position small/medium multi-robot systems as prime future candidates for outdoor UGVs in agriculture. Additionally, UGVs for smart farms should exhibit some quantitative physical characteristics founded on past developments and current studies that are summarized in **Table 5**.

Finally, autonomous robots of any type, working in fleets or alone, are essential for the precision application of herbicides and fertilizers. These activities reduce the use of chemicals generating important benefits: (i) a decrease in the cost of chemical usage, which impacts in the system productivity; (ii) an improvement in safety for operators, who are moved far from the vehicles; (iii) better health for the people around the fields, who are not exposed to the effects of chemical; and (iii) improved quality of foods that will reduce the content of toxic products.

Acknowledgements

The research leading to these results has received funding from (i) RoboCity2030-DIH-CM Madrid Robotics Digital Innovation Hub ("Robótica aplicada a la mejora de la calidad de vida de los ciudadanos. fase IV"; S2018/NMT-4331), funded by "Programas de Actividades I+D en la Comunidad de Madrid" and cofunded by Structural Funds of the EU; (ii) the Agencia Estatal Consejo Superior de Investigaciones Científicas (CSIC) under the BMCrop project, Ref. 201750E089; and (iii) the Spanish Ministry of Economy, Industry and Competitiveness under Grant DPI2017-84253-C2-1-R.

Author details

Pablo Gonzalez-De-Santos*, Roemi Fernández, Delia Sepúlveda, Eduardo Navas and Manuel Armada
Center for Automation and Robotics (UPM-CSIC), Madrid, Spain

*Address all correspondence to: pablo.gonzalez@car.upm-csic.es

References

[1] CEMA–European Agricultural Machinery. [Accessed: November 2019]

[2] Burke R, Mussomeli A, Laaper S, Hartigan M, Sniderman B. The Smart Factory: Responsive, adaptive, Connected Manufacturing. Deloitte University Press. Available from: https://dupress.deloitte.com/dup-us-en/focus/industry-4-0/smart-factory-connected-manufacturing.html, 2019 [Accessed: November 2019]

[3] Corke PI, Roberts JM, Winstanley GJ. Robotics for the mining industry. In: de Almeida A. T., Khatib O. (Eds) Autonomous Robotic Systems. Lecture Notes in Control and Information Sciences, Vol 236. Springer, London, 1998

[4] Bechar A, Vigneault C. Agricultural robots for field operations: Concepts and components. Biosystems Engineering. 2016;**149**:94-111

[5] Bechar A, Vigneault C. Agricultural robots for field operations. Part 2: Operations and systems. Biosystems Engineering. 2017;**153**:110-128

[6] Autopilot. Available from: http://www.trimble.com/Agriculture/autopilot.aspx, 2019 [Accessed: November 2019]

[7] AutoTrac. Available from: https://www.deere.com/en_INT/products/, 2019 [Accessed: November 2019]

[8] Lee EA, Seshia SA. Introduction to Embedded Systems - A Cyber-Physical Systems Approach. 2nd ed. Cambridge, Massachusetts: MIT Press; 2017

[9] Ochoa SF, Fortino G, Di Fatta G. Cyber-physical systems, internet of things and big data. Future Generation Computer Systems. 2017;**75**:82-84

[10] TMR, Transparency Market Research, Mobile Robotics Market: Global Industry Analysis, Size, Share, Growth, Trends and Forecast, 2017-2025. Available from: https://www.transparencymarketresearch.com/mobile-robotics-market-2017-2025.html, 2019 [Accessed: November 2019]

[11] CASE. 2019. Available from: https://www.therobotreport.com/case-ih-displays-new-cab-less-concept-tractor/, 2019 [Accessed: November 2019]

[12] NEW HOLLAND. Available from: http://agriculture1.newholland.com/nar/en-us/about-us/whats-up/news-events/2016/new-holland-nh-drive-new-concept-autonomous-tractor, 2019 [Accessed: November 2019]

[13] EU SCAR. Agricultural Knowledge and Innovation Systems Towards the Future – A Foresight Paper, Brussels. Luxembourg: Publications Office of the European Union; 2016

[14] Vetter AA. Quantitative evaluation of DGPS guidance for ground-based agricultural applications. Applied Engineering in Agriculture. 1995;**11**(3):459-464

[15] O'Connor M, Bell T, Elkaim G, Parkinson B. Automatic steering of farm vehicles using GPS. In: Proceedings of the 3rd International Conference on Precision Agriculture, Minneapolis, USA; 1996. pp. 767-777

[16] Noguchi N, Reid JF, Will J, Benson ER, Stombaugh TS. Vehicle Automation System Based on Multi-sensor Integration> ASAE Paper, 983111; 1998

[17] Pilarski T, Happold M, Pangels H, Ollis M, Fitzpatrick K, Stentz A. The Demeter system for automated harvesting. In: Proceedings of the

8th International Topical Meeting on Robotics and Remote Systems; 1999

[18] Stentz A, Dima C, Wellington C, Herman H, stager D. A system for semi-autonomous tractor operations. Autonomous Robots. 2002;**13**:87-104

[19] Subramanian V, Burks TF, Arroyo AA. Development of machine vision and laser radar based autonomous vehicle guidance systems for citrus grove navigation. Computers and Electronics in Agriculture. 2006;**53**(2):130-143

[20] Thuilot B, Cariou C, Martinet P, Berducat M. Automatic guidance of a farm tractor relying on a single CP-DGPS. Autonomous Robots. 2002;**13**(1):53-71. DOI: 10.1023/A:1015678121948

[21] Nørremark M, Griepentrog HW, Nielsen J, Søgaard HT. The development and assessment of the accuracy of an autonomous GPS-based system for intra-row mechanical weed control in row crops. Biosystems Engineering. 2008;**101**:396-410

[22] Gonzalez-de-Santos P, Ribeiro A, Fernandez-Quintanilla C, Lopez-Granados F, Brandstoetter M, Tomic S, et al. Fleets of robots for environmentally-safe Pest control in agriculture. Precision Agriculture. 2017;**18**(4):574-614

[23] Kayacan E, Kayacan E, Ramon H, Saeys W. Towards agrobots: Identification of the yaw dynamics and trajectory tracking of an autonomous tractor. Computers and. Electronics in Agriculture. 2015;**115**:78-87

[24] Bergerman M, Maeta SM, Zhang J, Freitas GM, Hamner M, Singh S, et al. Robot farmers: Autonomous orchard vehicles help tree fruit production. IEEE Robotics and Automation Magazine. 2015;**22**(1):54-63

[25] ISO, Robotic platform: ISO 500 ISO 730 ISO 11001 ISO/TR 12369— Agricultural tractors, 2017

[26] ATC - Autonomous Tractor Corporation – AutoDrive. 2019. Available from: https://www.autonomouStractor.com/ [Accessed: November 2019]

[27] Precision Makers, X-PERT. Available from: http://www.precisionmakers.com/en/x-pert/, 2019 [Accessed: November 2019]

[28] Garcia E, Gonzalez-de-Santos P. On the improvement of walking performance in natural environments by a compliant adaptive gait. IEEE Transactions on Robotics. 2006;**22**(6):1240-1253

[29] Lakkad S. Modeling and simulation of steering systems for autonomous vehicles. master thesis. The Florida State University, US; 2004

[30] Bak T, Jakobsen H. Agricultural robotic platform with four wheel steering for weed detection. Biosystems Engineering. 2004;**87**(2):125-136

[31] Bawden O, Ball D, Kulk J, Perez T, Russell R. A lightweight, modular robotic vehicle for the sustainable intensification of agriculture. In: Australian Conference on Robotics and Automation (ACRA 2014), University of Melbourne, December 2014

[32] Bawden O, Kulk J, Russell R, McCool C, English A, Dayoub F, et al. Robot for weed species plant-specific management. Journal of Field Robotics. 2017;**34**:1179-1199

[33] Bogue R. Robots poised to revolutionise agriculture. Industrial Robot International Journal. 2016;**43**(5):450-456

[34] Underwood JP, Calleija M, Taylor Z, Hung C, Nieto J, Fitch R, et al. Real-time

target detection and steerable spray for vegetable crops. In: Proceedings of IEEE ICRA -Workshop on Robotics in Agriculture; 2015

[35] Kongskilde, New Automated Agricultural Platform–Kongskilde Vibro Crop Robotti. Available from: http://conpleks.com/robotech/new-automated [Accessed: November 2019]

[36] Raussendorf, Raussendorf Maschinen. Available from: http://www.raussendorf.de/en/fruit-robot.html [Accessed: November 2019]

[37] Precision Makers, Greenbot. Available from: http://www.precisionmakers.com/greenbot/ [Accessed: November 2019]

[38] Gonzalez-de-Santos P, Garcia E, Estremera J. Quadrupedal Locomotion: An Introduction to the Control of Four-Legged Robots. London: Springer-Verlag; 2006

[39] Ruckelshausen A, Biber P, Dorna M, Gremmes H, Klose R, Linz A, et al. An autonomous field robot platform for individual plant phenotyping. Precision Agriculture. 2009;**9**:841-847

[40] Blackmore BS, Have H, Fountas SA. Specification of behavioural requirements for an autonomous tractor. In: Proceedings of the 6th International Symposium on Fruit, Nut and Vegetable Production Engineering conference, Germany, Institute für Agrartechnik Bornim e.V; 2001. pp. 25-36

[41] Jadeja Y, Modi K. Cloud computing - concepts, architecture and challenges. In: International Conference on Computing, Electronics and Electrical Technologies (ICCEET), Kumaracoil, India; 2012. pp. 877-880

[42] ISO, Robot safety system: ISO 14121—Safety of machinery— Risk assessment (Part 1- Principles and Part 2 - Practical guidance and examples of methods (Technical report)) prepared by the Technical Committee ISO/TC 199; 2017

[43] Wei Z. Real-time optical flow sensor design and its application to obstacle detection. PhD thesis. Brigham Young University. Provo, Utah, US; 2006

[44] Gonzalez-de-Soto M, Emmi L, Benavides C, Garcia I, Gonzalez-de-Santos P. Reducing air pollution with hybrid-powered robotic tractors for precision agriculture. Biosystems Engineering. 2016;**143**:79-994

[45] Perez-Ruiz M, Gonzalez-de-Santos P, Ribeiro A, Fernandez-Quintanilla C, Peruzzi A, Vieri M, et al. Highlights and preliminary results for autonomous crop protection. Computers and Electronics in Agriculture. 2015;**110**:150-161

[46] Emmi L, Gonzalez-de-Soto M, Pajares G, Gonzalez-de-Santos P. New trends in robotics for agriculture: Integration and assessment of a real fleet of robots. The Scientific World Journal. 2014;**2014**:1-21. 404059